Volume 5

Materials

Robin Kerrod

18428597

OXFORD
UNIVERSITY PRESS

Great Clarendon Street, Oxford OX2 6DP
198 Madison Avenue, New York, New York 10016

Oxford New York

Athens Auckland Bangkok Bogotá Buenos Aires Cape Town
Chennai Dar es Salaam Delhi Florence Hong Kong Istanbul Karachi
Kolkata Kuala Lumpur Madrid Melbourne Mexico City Mumbai Nairobi
Paris São Paulo Shanghai Singapore Taipei Tokyo Toronto Warsaw

with associated companies in Berlin Ibadan

Oxford is a registered trade mark of Oxford University Press

First published 2002

Library of Congress Cataloging-in-Publication Data available.
ISBN 0-19-521859-0 (single volume), 0-19-521906-6 (set)

1 3 5 7 9 10 8 6 4 2

Designed and typeset by Full Steam Ahead
Printed in Malaysia

Contents

EARTH'S RICHES

Every hour of every day oilfields around the world pump over 100 million gallons of crude oil, or petroleum, out of the ground. Oil is one of the most valuable resources, or substances we need to produce the materials we use in our everyday lives.

Oil is valuable for two reasons. One, it can be made into fuels, such as gasoline. In fact, it is one of our main energy resources, along with natural gas and coal. Two, oil can be converted into useful chemicals. From these chemicals, many different products can be made, such as plastics, paints, and pesticides.

Minerals are very important resources too. They are the substances that make up the rocks in the Earth's crust. In some places, there are concentrations, or deposits, of useful minerals, which can be obtained by mining.

The vital ores

By far the most valuable of these deposits are materials called ores. These are the minerals that we can process into metals. Our modern civilization could not exist without metals. They enable us to build the many machines used in industry, transportation and in our homes.

▶ Natural deposits of sulfur are often found around volcanic vents (openings). Sulfur is used to make many things, including gunpowder, insecticides, and fertilizers.

▼ This map shows where major deposits of mineral ores are found around the world. Some ores, such as iron, are found widely. Others, such as aluminium, are found only in a few places.

Key
- ▾ aluminum
- ▪ copper
- • gold
- ▪ iron
- ♦ silver

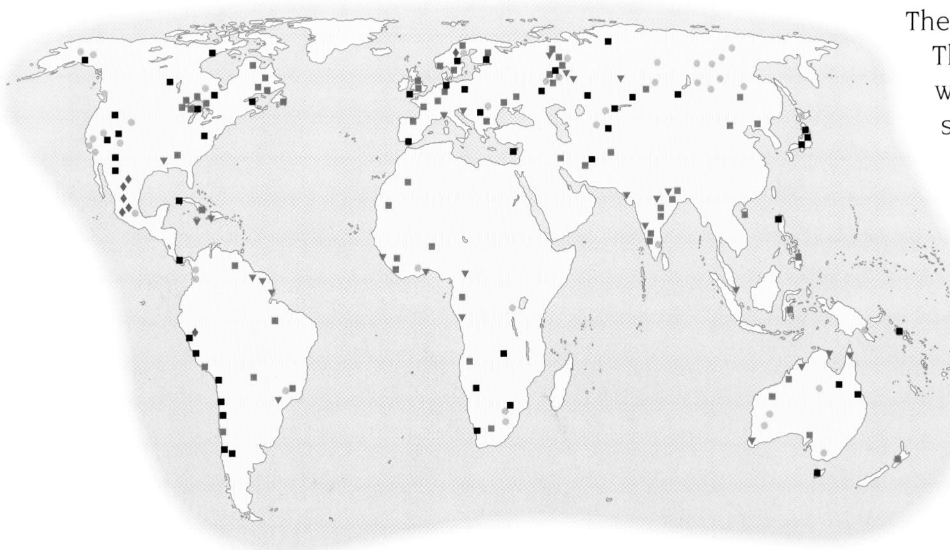

Other useful resources

Many other minerals and mineral mixtures are useful for industry. Limestone is used to make cement and in iron-making furnaces. Sand from lake shores and deserts is one of the main ingredients of glass. Clays are used to make bricks and pottery.

Many minerals can form crystals, which have regular shapes and flat surfaces. Some crystals are very beautiful and provide us with sparkling gems—diamonds, sapphires, rubies, and emeralds are the most prized.

Sea sources

The oceans are another valuable resource. They are full of minerals dissolved in the water. The main one is common salt, or sodium chloride. This is extracted from sea water on a large scale.

Sea water also contains magnesium salts. Most of the world's magnesium is extracted from sea water by a process known as electrolysis, which involves passing electricity through it.

Up in the air

Even the air we breathe is a useful source of chemicals for industry. Air is a mixture of many gases, the chief

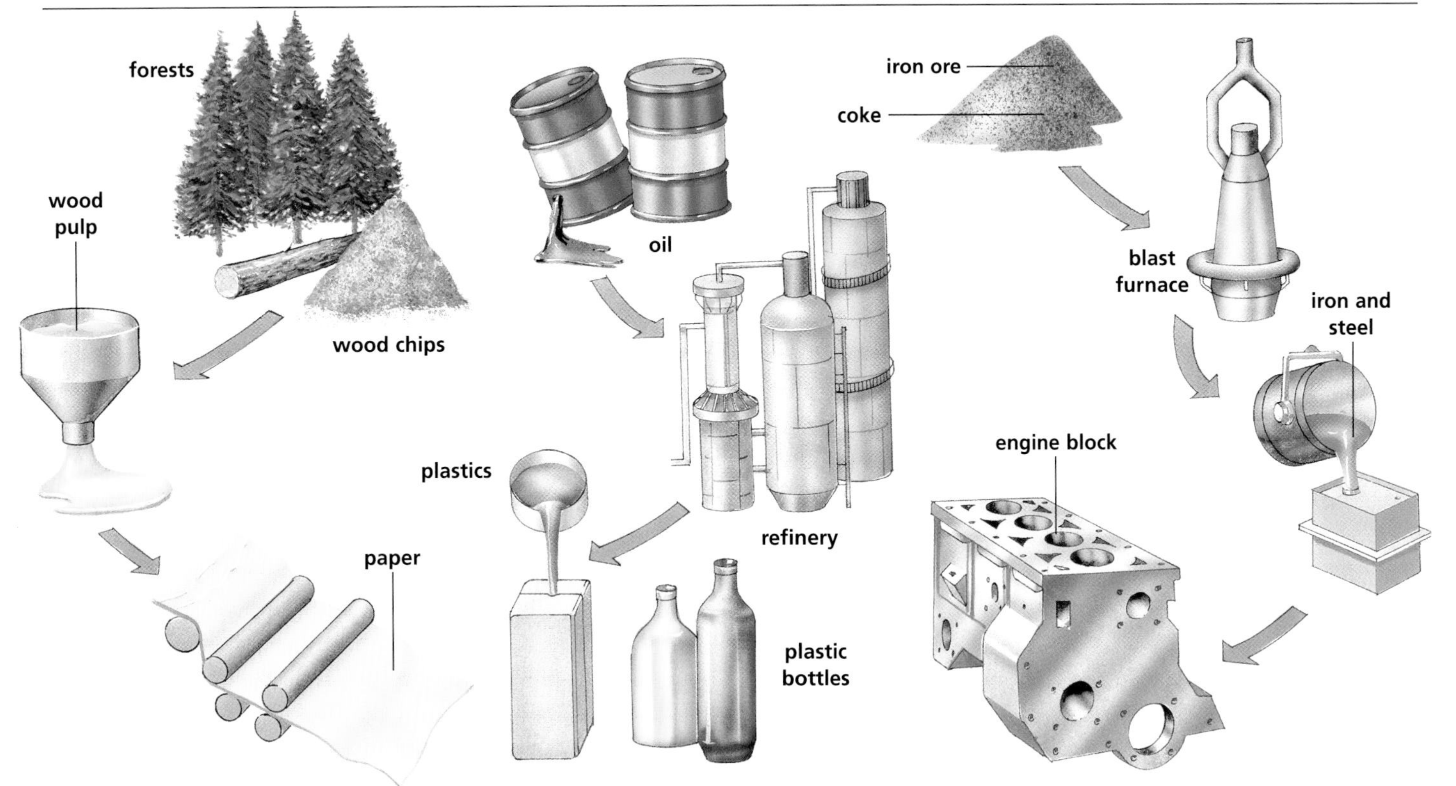

▲ Many of the products that we use every day have been made using natural resources.

ones being nitrogen and oxygen. Both these gases are widely used in industry. Nitrogen, for example, is used to make fertilizers,while liquid oxygen is used as a propellant in rockets.

The air is also the main source of the noble, or inert, gases. The most common of these gases, argon, is used to fill electric lightbulbs. Because argon is inert, or unreactive, it helps to prevent the bulb's white-hot wire filament from burning up.

Living resources

Living things can also be an important resource. Forests provide us with wood. Farms and plantations give us useful crops such as rubber and cotton. Other crops, such as sunflowers, flax, and canola are grown for the oil contained in their seeds.

Properly managed, forests and farms can continue to produce wood and crops year after year. They are known as renewable resources. Oil, coal, gas, and mineral deposits are not renewable. They took hundreds of millions of years to form, and once they have been used up, they will be gone for ever. This is why it is vitally important that we do our best to look after these precious resources.

key words

- minerals
- mining
- oil
- ores
- renewable resources
- salts

▼ Rain forests are rich in natural resources, including wood. By means of a process called photosynthesis, forest trees and plants also supply most of the world's oxygen.

MINING METHODS

In a tunnel nearly 2.4 miles (4,000 m) underground, the air is full of dust after an explosion. The temperature is 122°F (50 °C). Stripped to the waist, miners are loading shattered rock onto wagons. They are risking their lives to dig out one of the most precious metals on Earth—gold.

▶ In a silver mine in Mexico, miners drill holes in the rock, into which they will place explosives. The explosives will break up the rock, which contains the silver ore.

Many metals are mined, or dug out of the Earth's crust, underground. They include copper, zinc, lead, and nickel. These metals are found in minerals, or ores, in the rocks. To mine these ores, shafts are bored vertically down into the ground. Tunnels are then dug horizontally across to reach the ore deposits. Explosives are used to break up the rocks containing the ores, then the pieces are transported to the surface.

key words

- drilling
- opencut mine
- ores
- panning
- placer deposits
- quarrying

In most coal mines, a different method is used. Coal is found in thick layers, or seams, and is much softer than mineral ores. So it can be dug out by coal-cutting machines known as shearers.

Keeping cool

Large mines have hundreds of miles of tunnels on many levels, fanning out from many shafts. Some of the mines have lifts and railways to transport the miners to the ore deposits. Some shafts have hoists, or skips, to lift out the ore or coal. Other shafts are used for ventilation to supply fresh, cool air for the miners to breathe.

shaft mining
drift mining
hydraulic mining
opencut mining
air shaft
seams

◀ Different methods of mining are needed to reach mineral or coal deposits. Opencut mining digs out surface deposits. Hydraulic mining breaks up soft deposits with water jets. Underground deposits can be reached from hillsides or through shafts.

On the surface

It is cheaper and safer to mine on the surface of the ground. Fortunately, quite a few ore deposits are found on or near the surface. They include iron and copper ores and also the aluminium ore bauxite. Many coal seams are found near the surface too.

Mining at the surface is called opencut or strip mining. Mining begins by stripping off any soil, or overburden, covering the deposit. This is done by huge excavators, such as dragline scoops. If the ore or coal deposit is soft, it can be dug out and loaded into wagons or trucks. If it is hard, it must first be broken up by explosives.

Quarrying is the name given to the surface mining of rock, such as chalk, limestone, and marble. Marble is widely used for decoration in building work. It has to be removed carefully by driving wedges into natural cracks in the rock.

The Bingham Canyon copper mine in Utah is the world's largest man-made hole. Covering an area of 4.2 square miles (7 sq km), it is nearly .6 mile (1 km) deep.

Flash in the pan

Gold is found deep underground, and on the surface in "placer" deposits in stream and river beds. In the early days of gold mining, miners would "pan" for gold using a shallow pan. They would swirl material from the river bed around in the pan with some water. The lighter, gravelly material would wash away, leaving the heavier gold behind. Today, mechanical panning methods are used.

Tin ore (cassiterite) is also found in placer deposits, particularly in Malaysia. The ore is dug out by huge dredges. These are floating platforms which use a conveyor belt of buckets to dig up material from the sea or river bed.

▲ Cassiterite (tin ore) is dug out from opencast mines in Malaysia, as well as from placer deposits.

◀ Gold miners use hoses to suck up mud and gravel from the bed of the Madre de Dios River, Peru. The mixture is fed through a device called a sluice box. Heavy specks of gold fall to the bottom, while the lighter mud and gravel flow back into the river.

Out of a hole

Resources can also be taken from the ground by drilling. For example, holes are drilled down to deposits of crude oil, or petroleum, which are then piped back to the surface.

Borehole mining is used to extract underground salt deposits. Water is pumped into a hole bored into the solid salt. It dissolves the salt and is then pumped back to the surface as brine, a mixture of water and salt. Evaporating the brine recovers the salt. Sulfur can be mined in a similar way, using very hot water to melt the deposit.

MIGHTY METALS

Pure iron is quite a weak metal and is not particularly hard. But if you add a tiny amount of carbon to it, it becomes both strong and hard. It turns into the most useful metal we know—steel.

The world uses more iron (in the form of steel) than all the other metals put together. Yearly iron production is around 660 million tons, nearly 60 times as much as aluminum, our next most important metal.

▶ A blast furnace for making pig iron stands up to 195 feet (60 m) tall. A cart feeds iron ore, limestone, and coke into the top of the furnace. Molten iron collects at the bottom, with slag on top. The iron is further refined (purified) to steel in a basic oxygen converter. This uses a jet of oxygen to burn off carbon and other impurities.

The main use for steel is in construction. It is used to build bridges and skyscrapers and all kinds of machines and vehicles. Most tools are made from steel; so are most cans. Cutlery is usually made from stainless steel. This is one of many steel alloys, which contain a mix of other metals.

Extracting the iron

Iron is found throughout the world, but not in metal form. Instead, it is found as an ore —the iron is found within rocks, combined with other elements. There are several iron ores, including magnetite and hematite.

key words

- alloy
- blast furnace
- iron
- slag
- smelting
- steel

◀ Wearing protective clothing and using a long rod, a worker takes a sample of molten iron from a blast furnace.

Iron is extracted from iron ore by smelting—heating it to a high temperature, along with other materials. Smelting is carried out in a blast furnace, so called because hot air is blasted into it. The excess carbon quickly burns off, and the impurities form a slag.

Reducing the ore

Iron ore is fed into the blast furnace together with coke and limestone. The hot air being blasted in makes the coke burn fiercely, and temperatures rise as high as 2,900°F (1,600°C). As the coke burns, carbon monoxide gas is produced. This combines with the oxygen in the iron ore, leaving behind iron metal. At such a high temperature, the iron is molten (liquid) and trickles down to the bottom of the furnace.

Meanwhile, impurities in the ore combine with the limestone to form a molten slag. This is lighter and floats on top of the iron. From time to time, both the iron and the slag are removed. In this state, the iron is known as pig iron.

HENRY BESSEMER

The Romans were making a kind of steel over 2,000 years ago. But steel only began to be easily and cheaply made in 1856. In that year, the English industrialist Henry Bessemer (1813–1898) invented a way of producing steel by blowing air through molten pig iron to burn out the impurities. Bessemer's process led to the mass production of cheap, good-quality steel. This made possible new kinds of building and ship, and greatly improved the railways. The basic oxygen converter used in steelmaking today is a refinement of Bessemer's furnace.

▼ Red-hot molten steel is poured into a container. The steel has been produced using the basic oxygen process, which changes pig iron into steel with the help of a blast of oxygen.

The first iron people used was metal that fell to Earth from outer space, in meteorites.

When the molten iron leaves the furnace, it flows along a channel into molds. These molds are called pigs, because they are clustered around the channel like a group of suckling piglets around their mother.

Refining the iron

Pig iron still contains many impurities, especially excess carbon (from the coke). They must be removed before the metal can become really useful. Refining, or purifying, the metal, takes place in other furnaces.

Most pig iron is refined by the basic oxygen process. The iron is poured in its molten state into a conical vessel called a converter. Then a high-speed jet of oxygen is blasted into it. Most of the carbon quickly burns off, while other impurities form a slag. In a typical converter, up to 440 tons of pig iron can be converted to steel in about half an hour.

The best-quality steels, such as stainless steels, are made in electric-arc furnaces. Usually, these steels are made using steel scrap, rather than molten iron. The steel is heated to high temperature by means of an electric arc, a kind of giant electric spark.

MAKING LIGHT OF IT

Without aluminum, aircraft would probably still be built of wood, wire, and fabric. And we would probably not be living in a space age. But thanks to aluminum we can build planes and spacecraft that are both lightweight and strong.

After iron, alumnum is the second most important metal to us. Over 20 million tons of aluminum are produced each year. Its lightness is the main reason why aluminum is so useful. It is less than half as heavy as iron. Unlike iron, it does not corrode, or rust.

Pure aluminum is soft and weak. It is used as thin foil—for cooking, for example. But it becomes much more useful when traces of other metals (such as copper) are mixed with it to form alloys.

▶ Aluminum is used to make bodies for aircraft and other vehicles such as cars and buses.

key words

- alloy
- bauxite
- conductor
- corrosion
- electrolysis

Lightweight conductors

Aluminum alloys have many uses. The alloys from which aircraft and spacecraft are made are as strong as steel but very light. Some are used to make cookware for the home. They heat up readily because aluminum is a good conductor of heat, allowing heat to travel through it easily. The metal is also a good conductor of electricity, which is why it can be used for the transmission lines that carry electricity from power stations to our homes.

Making the metal

There is more aluminum in the Earth's crust than any other metal (8 percent). It is found combined with other elements in many minerals in clays and rocks. But it can only easily be extracted from an ore called bauxite, which contains the mineral alumina, or aluminum oxide. Australia has the largest bauxite deposits.

Aluminum metal is obtained from alumina by electrolysis—passing electricity through it. The process was discovered independently in 1886 by Charles Hall in the U.S. and Paul Héroult in France.

rotary kiln (dryer)
settling tank
dry alumina
bauxite
caustic soda
aluminum oxide (alumina)
electrolysis bath
filter
impurities
reaction vessel
pure aluminum
molten alumina and cryolite
electrodes

◀ To make aluminum, alumina (aluminum oxide) is first separated from other material in the bauxite ore. Electricity is then passed through a molten (liquid) mixture of alumina and cryolite (another aluminum mineral).

THE GREAT CONDUCTOR

Nearly 11 million tons of copper are produced worldwide every year. About half of it is used by the electrical industry because copper conducts, or passes on, electricity better than any other metal except silver. And silver is too expensive to use for electrical wiring.

Millions of miles of copper wires carry electricity into and around homes throughout the world. Copper is easy to make into wires because it is one of the most ductile of metals. This means that it can be pulled, or drawn out, into long lengths without breaking.

Copper is not only used by itself, but also in alloys, or mixtures with other metals. Common copper alloys are bronze, brass, and cupronickel. Copper and its alloys are very useful to us because they can resist corrosion, or rusting. This means that they last a long time.

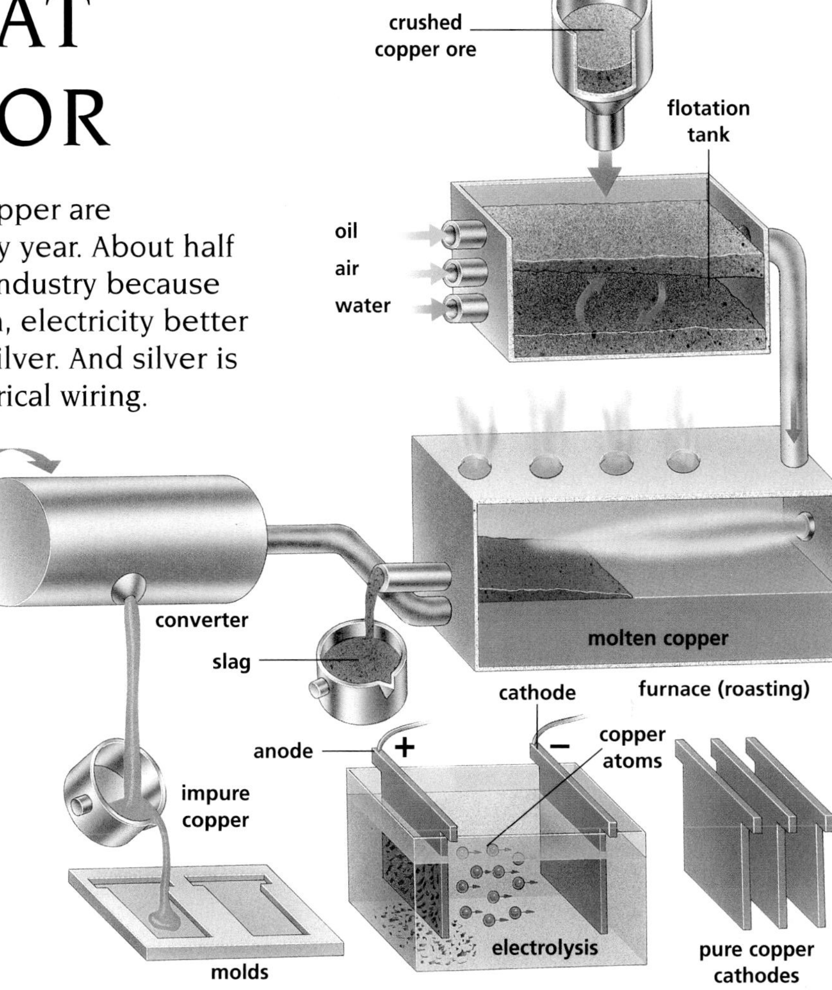

▲ To extract copper from copper ore, the ore is crushed and then fed to flotation tanks. These concentrate the copper minerals. The minerals are roasted in furnaces, where impurities burn off or form a slag. The nearly pure metal is finally refined by electrolysis. Electricity causes pure metal to be deposited on the cathode, or negative electrode, of the electrolysis cell.

▼ Bingham Canyon copper mine in Utah is the largest opencut mine in the world. It is nearly 0.6 mile deep and 2.4 miles wide. The mine is the richest source of copper in the world, but mining operations have caused a great deal of damage to the environment.

Going native

Copper was one of the first metals people used, at least 10,000 years ago. This is because copper is one of the few metals that can be found in metal form in the ground. We call this kind of metal "native metal."

But native copper is rare, and most copper is produced from copper ores found in rocks. There are huge deposits of copper ores in the Andes Mountains of South America, the Rocky Mountains of North America, and in Congo and Zambia, sometimes called the copper belt, in central Africa.

key words
- alloy
- conductor
- corrosion
- electrolysis
- native metal
- ore

PRECIOUS METALS

In 1939, a horde of treasure was found in the remains of a buried ship at Sutton Hoo in eastern England. It included 41 items made of solid gold, among them a beautiful helmet. When they were washed, they looked as good as they must have when they were buried 1,300 years earlier.

◀ This gold necklace, found in a Roman grave, is almost 2,000 years old.

Gold has always been admired for its rich beauty. It remains beautiful because it does not corrode, or rust, in the air or in the ground.

Because gold is so prized, it is called a precious metal. It was once used to make coins, but its main use today is in jewelry. Usually, other metals (such as copper) are added to it to make a harder alloy.

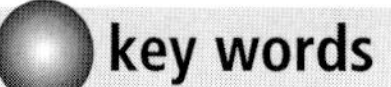

key words

- alloy
- corrode
- native metals
- ores

▶ This piece of rock, which contains native gold, comes from California.

Beat that!

Many metals can be beaten into shape. We say they are malleable. Gold is the most malleable metal of all. It can be beaten into gold leaf so thin that 10,000 sheets stacked together would measure only 0.04 inch (1 mm) thick. Gold is also the most ductile of metals, which means that it can be pulled into very fine wire without breaking.

Silver is another beautiful precious metal that has been used for jewelry since ancient times. But its main use today is in photography, because many silver compounds darken when they are exposed to light.

Like gold, silver is ductile and malleable. It is also the best conductor of heat and electricity that we know.

▼ When gold was discovered in Australia in the 19th century, people flocked to the goldfields from all over the world. This painting shows prospectors hoping to strike it rich.

The platinum group

Some metals are even more precious than gold. Platinum, for example, is rarer, harder, and melts at a higher temperature. Its main use in industry is as a catalyst, a substance that helps chemical reactions take place. Platinum is one of a group of heavy metals with similar properties.

Rich deposits

All the precious metals are found native, or in metal form in the ground. Native deposits provide most gold and platinum. But most silver comes from silver ores, such as argentite. The ores are treated with chemicals to produce the metal.

METAL MIXTURES

Copper and tin have many useful properties—for example, they do not easily corrode, or rust. The trouble is that they are quite soft and weak. But the metal you get when you mix them together is hard and strong, and still does not rust. It is one of our most useful metal mixtures, or alloys, and it is called bronze.

Like copper and tin, most metals are quite soft and weak in their pure state. But they become much harder and stronger when they are mixed with other metals to form alloys. Most metals used today are alloys, from the coins in our pockets and the cutlery we eat with, to the frames and engines of aircraft.

The right recipe

Alloys have been used for thousands of years. Today, metallurgists, the scientists who work with metals, can produce alloys with a wide range of different properties. They do so by carefully selecting the alloying ingredients—choosing the right "recipe."

Our most common metal, steel, is an alloy. It is a mixture of iron, a little carbon, and traces of other metals. It is the carbon that makes it so strong.

One drawback with ordinary steel is that it corrodes, or rusts. To stop it rusting, metallurgists add chromium and nickel to it. These metals do not rust and make the steel rust-resistant too. The alloy formed is stainless steel.

Some of the most advanced alloys are found inside jet engines. They have to remain strong at temperatures up to 1,830 °F (1,000 °C). Some of these so-called superalloys may contain 10 or more different metals, including titanium and tungsten, which melt only at high temperatures.

The value of alloys was discovered in very ancient times. Bronze was made before 3,000 B.C., and brass (copper and zinc) has been used for nearly as long. Pewter is an ancient alloy made from tin and lead, first used over 2,000 years ago. Modern pewter uses other metals instead of lead. This makes it safe to use for plates and mugs.

▶ Bronze is hard and does not rust. People have used this alloy for thousands of years to make things like sculptures.

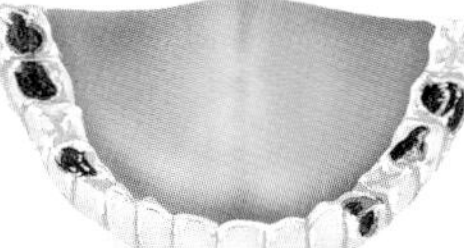

tool steel (iron, chromium, tungsten)

dental amalgam (mercury, silver, tin, zinc, copper)

duralumin (aluminum, copper, magnesium)

stainless steel (iron, chromium, nickel)

brass (copper, zinc)

cupronickel (copper, nickel)

▶ Some common alloys, with typical uses.

key words

- alloy
- corrode
- metallurgist

Memory metals are alloys of titanium and nickel that can "remember" their shape. If an object made of memory metal becomes twisted, it will return to its original shape if it is gently heated.

WORKING WITH METALS

The blacksmith is heating a strip of iron on a fierce fire. A horse waits patiently nearby. When the iron is red-hot, the blacksmith places it on an anvil. Then he begins hammering it into the shape of a horseshoe. He is carrying out the earliest method of shaping metal, called forging.

▶ A red-hot sheet of metal is rolled in a steel mill. Hot rolling is usually followed by cold rolling, which improves the surface finish.

Forging is still one of the main ways in which metal is shaped. But these days, forging is carried out in factories, using mechanical hammers. They have a heavy ram that shapes metal when it drops onto it from a height. This method is called drop forging. Usually, the metal is hammered into a shaped mold, or die. In an alternative method using a forging press, metal is shaped by a squeezing action rather than by sudden blows.

Molding metal

Another ancient way of shaping metal is by casting. The metal is heated until it is molten (liquid). Then it is poured into a shaped mold. It takes the shape of the mold when it cools and sets hard. Many castings are produced using molds made from wet sand. Mold-makers can make large and complicated shapes, such as ships' propellers. Sand molds have to be broken up to release the castings.

Many objects, however, are made in permanent molds called dies, which can be used over and over again. This method,

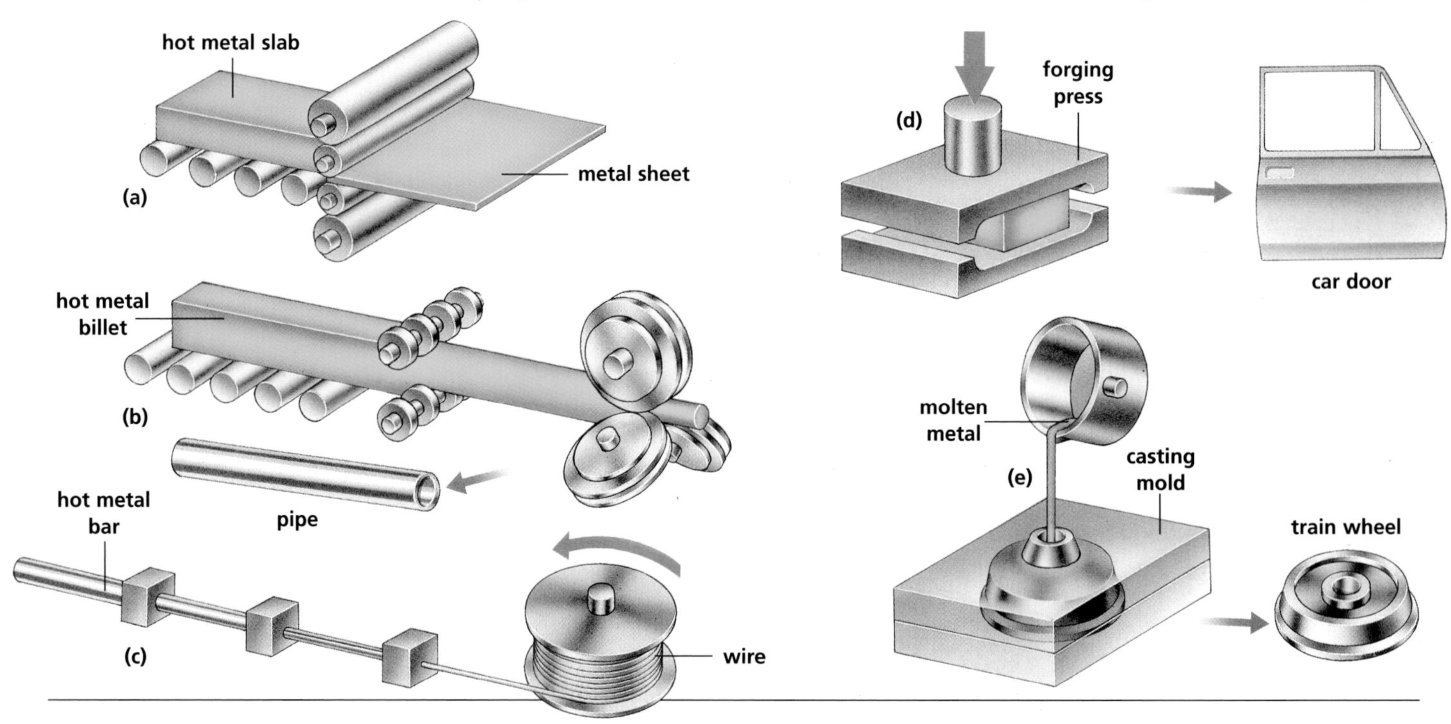

▼ Metal can be shaped in many different ways, including rolling (a, b), drawing (c), forging or pressing (d), and casting (e).

known as die-casting, is widely used for mass-producing small castings for toys and domestic appliances.

Rolling along

Huge amounts of metal are used in the form of plates or sheets, to make things like ships' hulls and car bodies. These products start life as thick slabs, which are made thinner and longer by rolling.

This is done in rolling mills. Here, red-hot slabs are passed back and forth through sets of heavy rollers. As the metal passes through each set, it gets thinner and longer. A thick slab originally 33 feet (10 m) long would typically end up as a sheet 0.08 inch thick and 0.9 mile (1.5 km) long.

After hot rolling, sheet steel is often rolled cold. This is done to give it an accurate thickness and a harder finish.

Pressing and stamping

One main use for sheet steel is for car bodies. The sheet is shaped cold on hydraulic presses, similar to but smaller than forging presses. Smaller versions of the drop forge are used to shape small objects, such as coins, from cold metal. They force metal into shape in dies, a process called stamping.

Other methods use dies for shaping. Rods and tubes may be made by extrusion, which involves forcing metal through holes in dies. Wire is made by drawing metal through sets of dies with smaller and smaller holes.

key words

- casting
- drawing
- forging
- pressing
- rolling
- soldering
- welding

▶ A huge hydraulic press is used to make a shaft for a steam turbine (a kind of engine used in power stations). The press can exert a force of 4,400 tons on the hot metal.

Electrical arcs (large sparks) can form underwater as well as in air. This means that electric arc welding can be used to join pieces of metal underwater.

▼ A welder joins pieces of metal that have been heated and softened using an electric arc.

Joining up

Metal pieces often need to be joined together to make large objects, such as ships' hulls and pipelines. Welding is the most common method.

In welding, the edges of the parts to be joined are first softened by heating them to high temperatures. Then molten metal from a filler rod is added. The added metal bonds with the softened metal. This fills in the gaps and produces a strong joint when it cools. Gas welding uses a burning gas torch to heat the metal. Arc welding uses an electric arc to heat it.

Soldering is another method using molten metal (solder) to form joints. It is used mostly to join wires in electrical circuits. Soldering is carried out at much lower temperatures than welding, and joints are not as strong.

BAKED EARTH

When the space shuttle returns to Earth, it uses the atmosphere to slow it down. Friction of the air against the fast-moving craft produces great heat, which makes the tiles covering the shuttle glow red-hot. But the astronauts inside are safe because the tiles stop the heat from reaching them.

The space shuttle tiles are products we call ceramics. Many ceramics are made by baking earthy materials such as sand and clay at high temperatures. Pottery is the most common ceramic product, and has been around for many thousands of years. So have two other common ceramic materials—bricks and tiles.

Preparing pottery

All pottery is made from clay. Different kinds of pottery are made from different kinds of clay and are fired (baked) at different temperatures. Firing takes place in ovens called kilns.

The ordinary kind of pottery, such as the crockery we use everyday, is known as earthenware. It is fired at about 1,800°F (1,000°C).

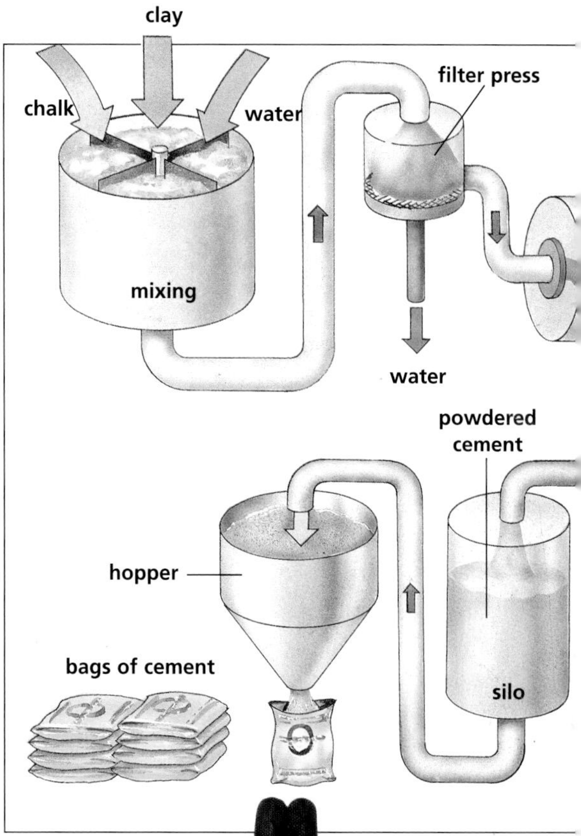

▶ An ancient Greek black-figure vase showing women running. In this kind of vase, the figures were painted in black. Details were added to the black figures by scraping away the paint to reveal the red clay beneath.

▼ A potter shapes wet clay with his hands as it spins round on a rotating wheel.

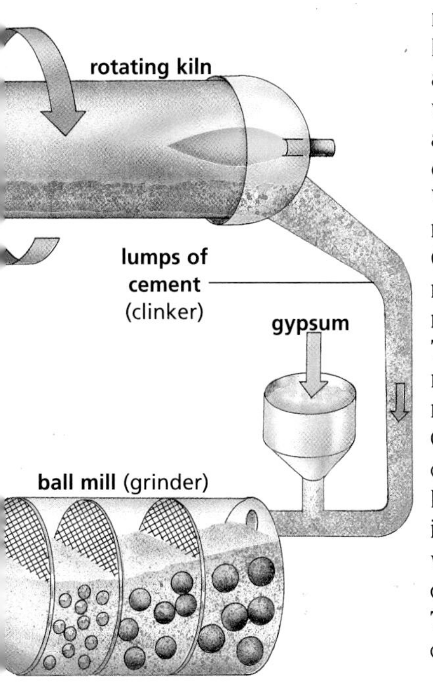

CEMENT AND CONCRETE

Concrete is the most widely used material in engineering construction. Hundreds of thousands of tons of it are used every day throughout the world. It is made by adding water to a mixture of cement, sand, and gravel or stones, to form a pasty mass. When the mass sets, it forms a material that is very hard—concrete. Cement is a ceramic product, made by fiercely heating earthy materials like limestone and clay. This is usually done in a long, rotating kiln in which temperatures may reach 2,700°F (see left). Concrete is strong under compression—when it is squeezed. But it is weak under tension—when it is stretched. To prevent this weakness, construction engineers cast steel rods into the concrete. The result is called reinforced concrete.

By itself it is dull in appearance and porous, which means that it lets water through. To make it look better and make it watertight, it has to be glazed, which involves giving it a glassy coating.

The finest-quality pottery is porcelain. This is made from only the purest white clays, such as kaolin, also called china clay. It is fired at temperatures up to about 2,550°F (1,400°C). At these temperatures, the clay vitrifies, or becomes glasslike. This makes it watertight. Bone china is an imitation porcelain made using clay mixed with bone ash.

Brickmaking

The first bricks were shaped blocks of mud that were dried in the sun. Houses are still built of mud bricks, called adobe, in some sunny countries.

Ordinary house bricks are now made of a mixture of clay, shale, and iron ore. The mixture is first crushed fine and then kneaded with water into a doughy mass. This is forced through an opening to form a long ribbon, rather like toothpaste being squeezed out of a tube. Rotating wires cut the ribbon into individual bricks. The bricks then pass through a tunnel-like kiln, which slowly heats them, then cools them again.

key words

- ceramics
- clay
- firing
- kiln
- refractory

The wheel was first used for making pottery around 3500 B.C., probably before it was used to make the first vehicles.

Remarkable refractories

Special bricks are made to line the inside of industrial furnaces, like those used to make iron and steel. They are made from naturally occurring minerals such as silica (sand), dolomite, and alumina, which melt only at high temperatures.

▶ A technologist removes a sample of cermet from a furnace. This cermet (ceramic metal) is made by heating and mixing together the ceramic material boron carbide and aluminum. It is lighter than aluminum and stronger than steel.

Materials that resist high temperatures are known as refractories. They include the space shuttle tiles already mentioned, which are made out of silica fibers. Some of the best refractory materials contain tungsten, the metal with the highest melting point 6,170°F (3,410°C). They include tungsten carbide, which is used to make cutting tools that remain sharp even when they get red-hot. Tungsten and titanium carbides are mixed with ceramics to form cermets, which are used in the high-temperature parts of jet and rocket engines.

LOOKING AT GLASS

Take some sand and some limestone, two of the most common materials in the ground. Add some soda ash, and heat the mixture to about 2,730°F (1,500°C), until it becomes a red-hot, molten mass. Let it cool, and the mass becomes transparent. It has turned into glass.

Glass is one of the most remarkable materials there is. Not only is it transparent, but it is also waterproof and does not rot or rust. It is resistant to all common chemicals and is easy to clean. It can easily be shaped into blocks, sheets and fibers. Glass fibers are widely used for reinforcing (strengthening) plastics. In fiber optics, very fine glass fibers are used to carry data and communications, such as telephone calls.

The main ingredient in glass-making is sand, the mineral silica. If this is melted, then cooled, it forms glass. But other ingredients must be added to it to make it melt at a reasonable temperature. The ordinary glass used for bottles and windows is known as soda-lime glass because it is made using soda ash and limestone.

▶ A magnificent stained-glass window in Canterbury cathedral, England. Colored glasses are made by including compounds of certain metals (copper, chromium, nickel, cobalt) in the glass-making recipe.

key words

- borosilicate glass
- float glass
- lead glass
- silica

Special glasses

Other ingredients are added to the glass-making recipe to produce glasses with special properties. Adding lead oxide, for example, makes lead, or crystal, glass. This has extra brilliance and, when expertly cut, gleams and sparkles like diamond. Glass with a very high lead content is made for the nuclear industry, because it blocks harmful radiation.

Ordinary glass expands rapidly when it is heated. When you pour boiling water into a cold glass bottle, for example, the sudden expansion will make it crack and shatter. But when you add boron to the glass-making recipe, you produce a glass that hardly expands at all. This borosilicate glass is used to make heat-resistant cookware and laboratory equipment.

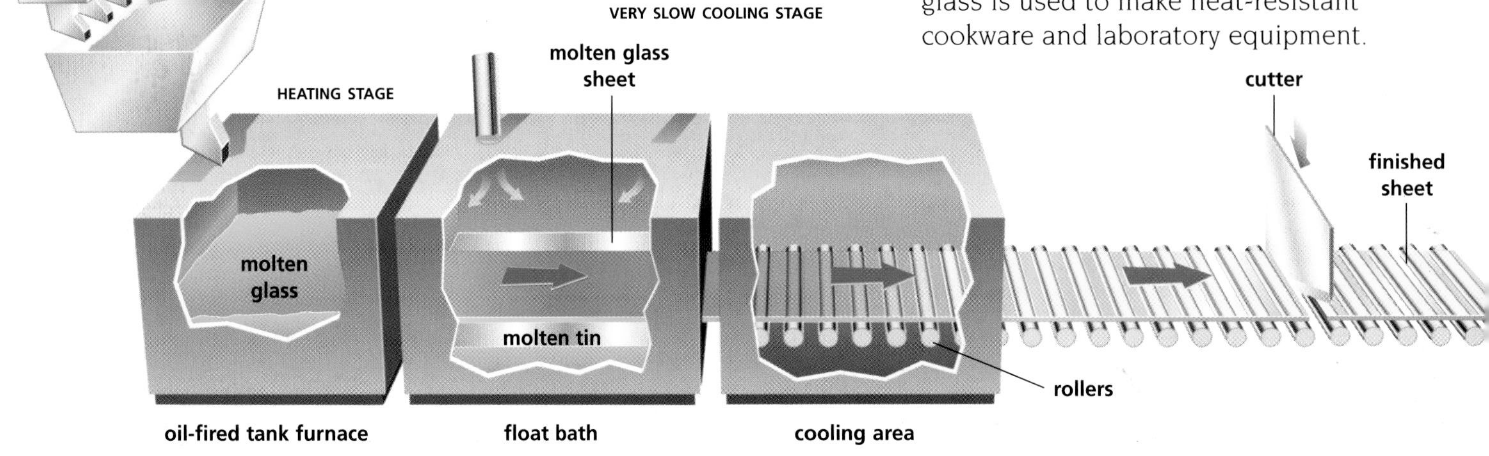

▼ To make sheets of float glass, a thin layer of molten (liquid) glass from the furnace is floated on a bath of molten tin. Because the surface of the liquid tin is perfectly flat, the glass layer is perfectly flat too.

FOREST PRODUCTS

In a forest in North America, the trees have been growing for more than a century—many are over 195 feet (60 m) tall. Now the buzzing of chainsaws signals that the lumberjacks have moved in. The chainsaws cut through thick trunks in minutes, and the mighty trees start crashing down.

Lumber—the wood from cut-down, or felled, trees—has always been one of the most useful materials to people. It is used in housebuilding, for making furniture, building boats, and much more besides.

Lumber is also used to make such products as plywood (thin layers of wood glued together to make boards) and chipboard (a material made from wood chips pressed and glued together). Wood is also a useful raw material for making such products as paper, textile fibers and even explosives. However, the biggest use of wood by far throughout the world is for fires for heating and cooking.

▼ Violins and many other musical instruments are usually crafted from the finest-quality hardwood.

Softwoods and hardwoods

The common lumber used in building is called softwood, because it is relatively soft and easy to cut and work with. Most softwood trees have narrow, needlelike leaves and bear their seeds in cones. They keep their leaves all year round. Known as evergreen conifers, they include firs, pines, cedars, and spruces. Large areas of natural conifer forests are found in the cool, northern regions of the world, and they are planted in warmer climates too.

Key

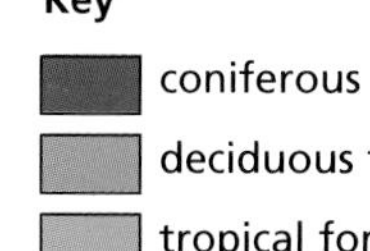

▲ The forest regions of the world. The largest region of natural forest is in the north of North America, Europe, and Asia. Evergreen conifers grow there. The other main area is around the Equator. In these tropical forests, broad-leaved evergreens grow. In the temperate climates between these main forest regions, the native trees are broad-leaved and deciduous.

However, in warmer regions the natural forest trees yield a harder wood, so they are called hardwoods. They include oak, beech, chestnut, and maple. These trees are deciduous, which means that they shed their broad leaves every autumn. Hardwood trees grow in the natural forests around the Equator. These broad-leaved trees are evergreen. The most valuable hardwoods include mahogany, ebony, and teak.

Managing the forests

At one time about two-thirds of the Earth's surface was covered by forests. But for

▲ In a tropical forest in Cameroon in central Africa, two men saw a massive tree trunk into logs.

thousands of years people have been felling trees for lumber, for firewood, and to clear land for agriculture. Only a fraction of the original forest land remains. And destruction of the natural forest continues, especially the tropical rainforests of Southeast Asia.

Elsewhere, more and more lumber is coming from managed forests. In these forests, the trees are cultivated as a crop, with new trees being planted as mature ones are cut down. Forestry workers raise tree seedlings, plant them, and look after them while they grow, protecting them from pests and diseases.

From forest to sawmill

To fell a tree, a lumberjack first cuts a wedge out of the trunk low down. Then he makes a slice above it. Because of the undercut wedge, the tree loses balance and topples over. After trees are felled, they are cut into logs. The logs may be taken out of the forest by tractors, animals (such as elephants), cables, or water slides. They travel to sawmills on trucks or railway flatcars, or are towed on huge rafts across lakes.

At the sawmill, the logs are debarked and then sawn into pieces of standard sizes. Afterward, the cut wood has to be stacked so that the air can dry it out. This is called seasoning. Sometimes the wood is seasoned artificially in heated kilns (ovens).

key words

- conifers
- deciduous
- evergreen
- felling
- seasoning
- lumber

Chemical products

Wood is made up mainly of fibers of cellulose, which are held together by a substance called lignin. Cellulose is the starting point for making rayon fibers, cellulose plastics, and explosives.

Useful oils and solvents can be obtained by distilling wood, which involves heating it in enclosed vessels. These products include creosote, used to preserve wood, and turpentine, used in paints. Partly burning wood produces charcoal, a valuable fuel.

▼ Many things that we use every day are made from wood or wood products.

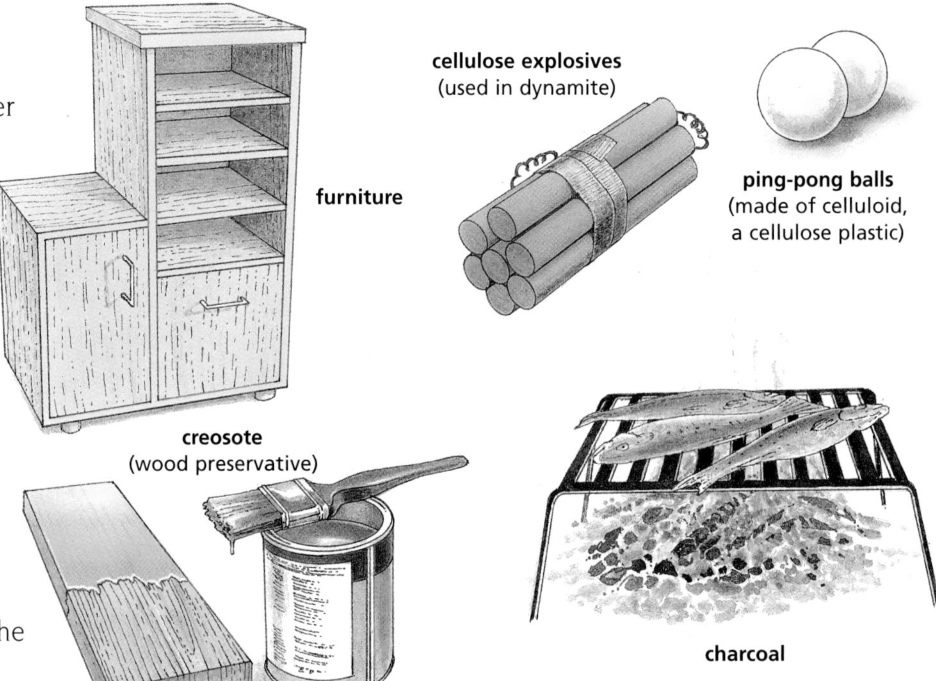

THIN SHEETS

As you read these words, you are looking at a material that started life in a forest in the wilds of Canada or Scandinavia. You are looking at paper, which is made from the wood of trees.

Paper is one of the greatest inventions of all time. It has allowed ideas and knowledge to be written down and passed on from generation to generation. Even in this electronic age, we are using more paper than ever before. Each year an area of forest the size of Sweden has to be cut down for papermaking.

▶ A worker in a paper mill takes a sample of wood pulp from a roller. The sample will be tested for quality.

◀ In the paper mill, logs are first chipped, then mixed with chemicals to make pulp. The pulp then goes to the paper-making machine. Excess water drains away, then the damp paper is pressed, dried and further rolled to smooth the surface.

Pulping

The first stage of papermaking is to turn the wood from logs into a mass of fibers, called pulp. The cheapest pulp is made by shredding the wood using a rotating grindstone. It is used to make newsprint, the paper newspapers are printed on. Better-quality pulp is made by treating wood chips with chemicals. The chemicals release the wood fibers (cellulose) by dissolving the substance that binds them together (lignin).

key words

- cellulose
- fibers
- lignin
- newsprint
- pulp
- wood

Papermaking

At the paper mill, the wood pulp first goes to a machine that beats the fibers and makes them frayed and flexible. The beaten pulp then goes into a mixer, where certain materials are added. They may include china clay to give the paper more weight; a gluelike material to make the paper easier to write on, and pigments to add color. The thoroughly mixed pulp is then fed to the papermaking machine.

FANTASTIC ELASTIC

The daredevil girl dives off the high bridge and plunges toward the river far below. Down she goes until she is only a few feet above the water. But she does not plunge in because she is bungee-jumping. An elastic rope attached to her feet pulls her back up.

The elastic rope is made of rubber. Being elastic is only one of rubber's useful properties. It is also airtight and waterproof, and absorbs shock well. These properties make it suited to two of its main uses—for car tires and footwear.

Rubber can be natural or synthetic. Both natural and synthetic rubber are made up of long molecules, which are folded back on themselves rather like a spring. They unfold when the rubber stretches, then spring back when released.

key words

- elastic
- latex
- polymers
- synthetic rubber
- vulcanization

▲ Once the different parts of a tire have been assembled, they are fused together, and the rubber is hardened, in a tire press. Here, a tire press releases a new tire, still steaming, from its mold.

Making rubber

Natural rubber is prepared from the white milky sap, or latex, of rubber trees. Synthetic rubber is made from chemicals obtained from petroleum, or crude oil. These chemicals form long chains, or polymers, when they react together, producing a substance similar to natural latex.

Some latex is made into products such as rubber gloves, but most is processed further. One of the most important processes is called vulcanization. This involves adding sulfur and heating the rubber. This makes it harder and tougher.

▼ Making natural rubber, and its use in manufacturing car tires.

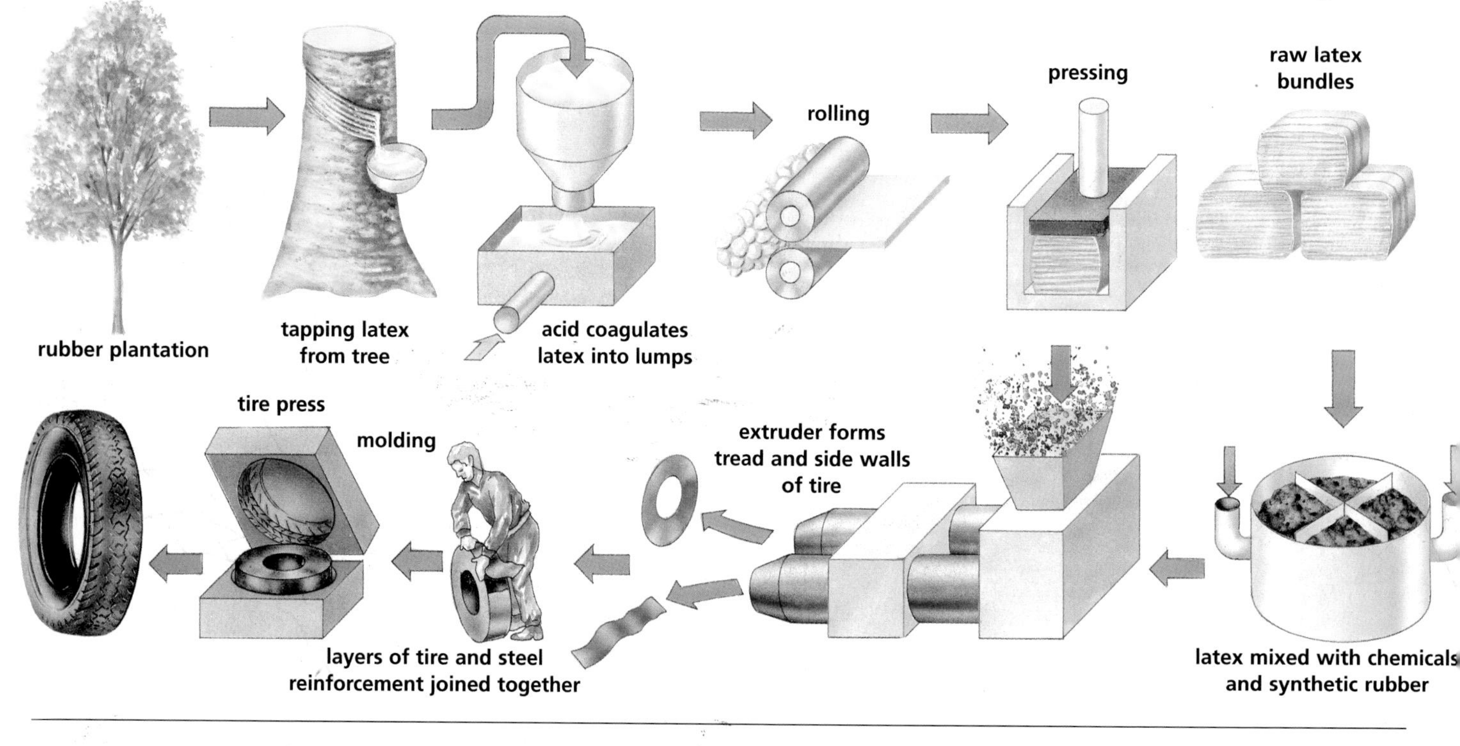

FIBERS AND FABRICS

The little worm has been gorging itself on mulberry leaves for five weeks. Now it is starting to spin its cocoon. Two streams of liquid coming from its spinning glands harden into fine threads as they hit the air. These threads are one of our most prized textile fibers—silk.

Silk is just one of several natural fibers we use to make fabrics, or textiles. Others include cotton, wool, and linen. Equally important these days are man-made fibers. Some, such as rayon, are made by processing natural materials. Others, called synthetic fibers, are made from chemicals.

Most textiles are made in two stages. First, bundles of fibers are drawn out and twisted into long threads, or yarns. This process is called spinning. Then the yarns are interlaced together, in a process called weaving.

▶ The thick fleeces being shorn from these sheep will be spun into woolen yarn to make clothing.

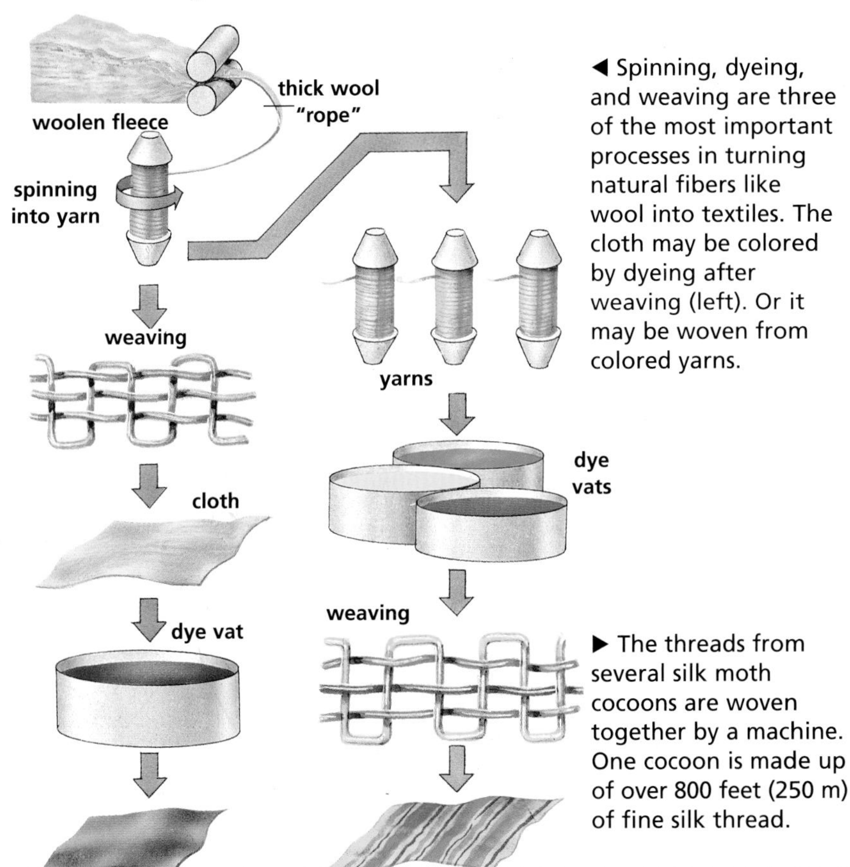

◀ Spinning, dyeing, and weaving are three of the most important processes in turning natural fibers like wool into textiles. The cloth may be colored by dyeing after weaving (left). Or it may be woven from colored yarns.

Animal fibers

Silk is one of two main fibers we get from animals. It is the only one that is produced in the form of a continuous thread, or filament. Several threads are twisted together to make a silk yarn strong enough for weaving.

The wool from sheep is the other main animal fiber. Australia is the world's biggest wool producer, with an output of about 1.7 million tons a year. Goats, camels, and llamas also produce useful fibers.

▶ The threads from several silk moth cocoons are woven together by a machine. One cocoon is made up of over 800 feet (250 m) of fine silk thread.

Plant fibers

Cotton is by far the most important plant fiber. Cotton plants produce fibers in the seed pod, or boll. They grow only in tropical and subtropical regions, with China and the southern states of the U.S. being leading producers. World output is about 13 million tons a year.

Linen is made from another plant fiber, the flax plant. These plants grow in cooler climates and are cultivated in much the same way as cereal crops. The fibers are found inside the stalks.

Rayon

Cotton is used as a raw material to make rayon, the most widely used man-made fiber. Wood pulp is also used to make rayon. Cotton and wood pulp both contain cellulose.

The most common form of rayon is called viscose. It is made by dissolving the cellulose with chemicals. This makes a syrupy solution, which is then pumped through tiny holes in a device called a spinneret, into a bath of acid. In contact with the acid, the streams of solution turn into continuous threads, or filaments, of pure cellulose.

key words

- cellulose
- fibers
- spinneret
- spinning
- synthetic
- weaving

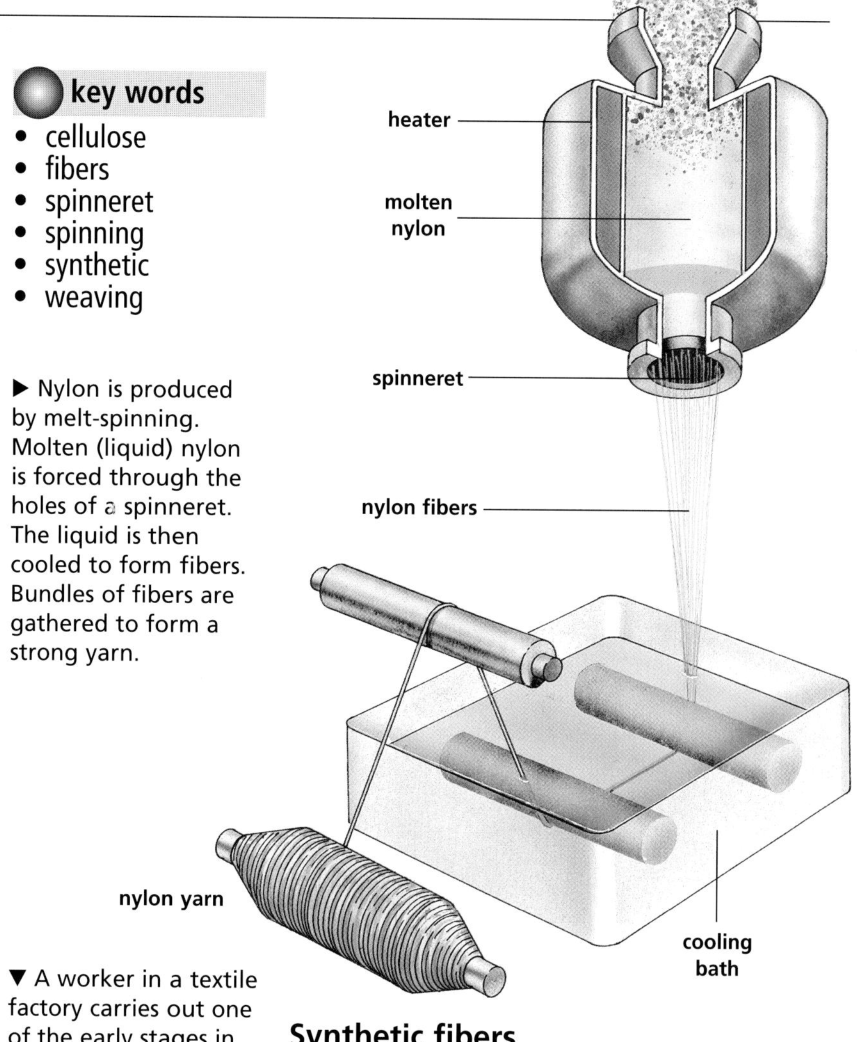

▶ Nylon is produced by melt-spinning. Molten (liquid) nylon is forced through the holes of a spinneret. The liquid is then cooled to form fibers. Bundles of fibers are gathered to form a strong yarn.

▼ A worker in a textile factory carries out one of the early stages in the weaving process.

Synthetic fibers

Synthetic fibers are kinds of plastics that can be pulled, or drawn out, into fine threads. They are made mainly from petroleum chemicals. The best-known synthetic fiber is nylon. Nylon was first made in 1935 by a team led by the American chemist Wallace Carothers.

Synthetic fibers are formed by so-called spinning processes. The plastic material is pumped through the holes of a spinneret and comes out as long continuous threads. The threads are usually chopped into shorter fibers, often mixed with natural fibers, and then spun into yarn for weaving.

Drip dry

Synthetic fibers have many advantages over natural fibers. They are usually much stronger, resist insect attack, and do not rot. They do not absorb water, which means that they dry quickly. They resist creasing too, so fabrics keep their shape well.

CHEMICAL CHANGES

Salt is a very useful substance. In the home, we use it to make our food taste better. In industry, it is a valuable raw material. From salt, a wide range of chemicals is produced, including sodium carbonate (used in making glass) and caustic soda (used in making soap).

The chemical industry makes millions of tons of sodium carbonate and caustic soda every year. They are produced in such large quantities that they are known as heavy chemicals. Other heavy chemicals include sulfuric and nitric acids, ammonia, and benzene.

key words

- catalyst
- chemical engineering
- molecules
- reactions
- sulfuric acid
- synthetic

▼ Froth flotation tanks like this are used to concentrate silver and zinc ores. Froth flotation is a process that separates the minerals from earthy impurities. The ore is ground to a fine powder and mixed with water and chemicals to make a froth liquid. The mineral particles cling to the froth bubbles at the surface, while waste earthy material sinks to the bottom.

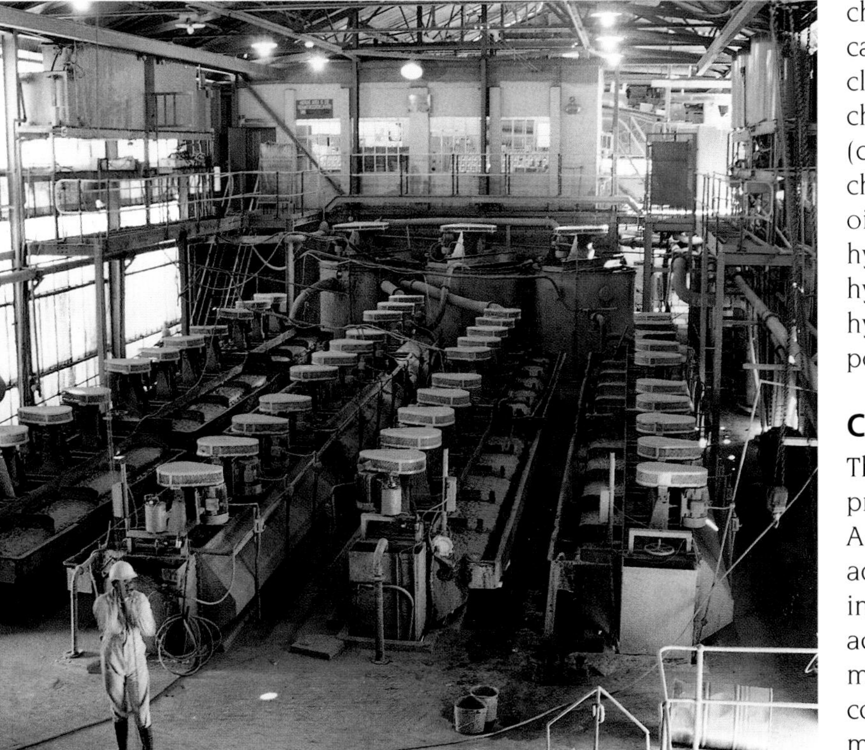

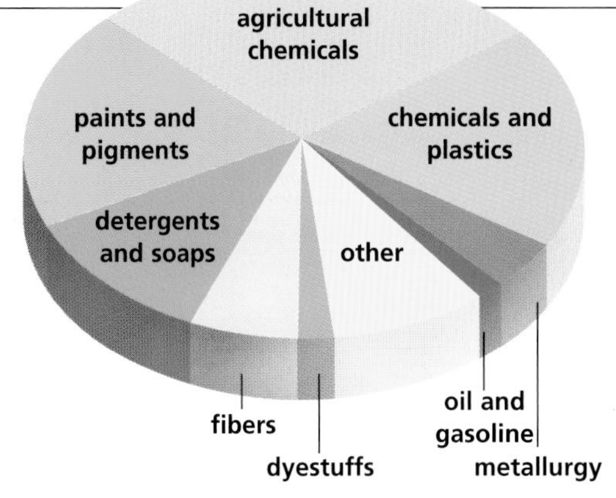

▲ The main uses of sulfuric acid.

SULFURIC ACID

The world produces almost twice as much sulfuric acid as any other chemical. It is so important in manufacturing that it is often called the "lifeblood of industry." Sulfuric acid is made from sulfur by the contact process. The sulfur (S) is first burned in air (containing oxygen, O_2) in a furnace, which produces sulfur dioxide (SO_2) gas. The gas then passes with more air into a converter containing a catalyst, such as platinum.
In contact with the catalyst, the sulfur dioxide combines with more oxygen from the air to form sulfur trioxide (SO_3). Sulfur trioxide can then be combined with water (H_2O) to form sulfuric acid (H_2SO_4).

Benzene is what is known as an organic chemical (organic chemicals contain carbon). The other heavy chemicals are classed as inorganic. Most inorganic chemicals are produced from minerals (chemicals extracted from rocks). Organic chemicals are mainly produced from crude oil, or petroleum, which is a mixture of hydrocarbons (compounds made from hydrogen and carbon). The different hydrocarbons provide chemical starting points for a huge variety of products.

Chemical operations

The chemical industry uses all kinds of processes, or reactions, to make chemicals. A common one is oxidation, which means adding oxygen to a substance. Oxidation is involved in the manufacture of sulfuric acid, for example. Hydrogenation, which means adding hydrogen, is another common reaction. It is used in making margarine and other spreads.

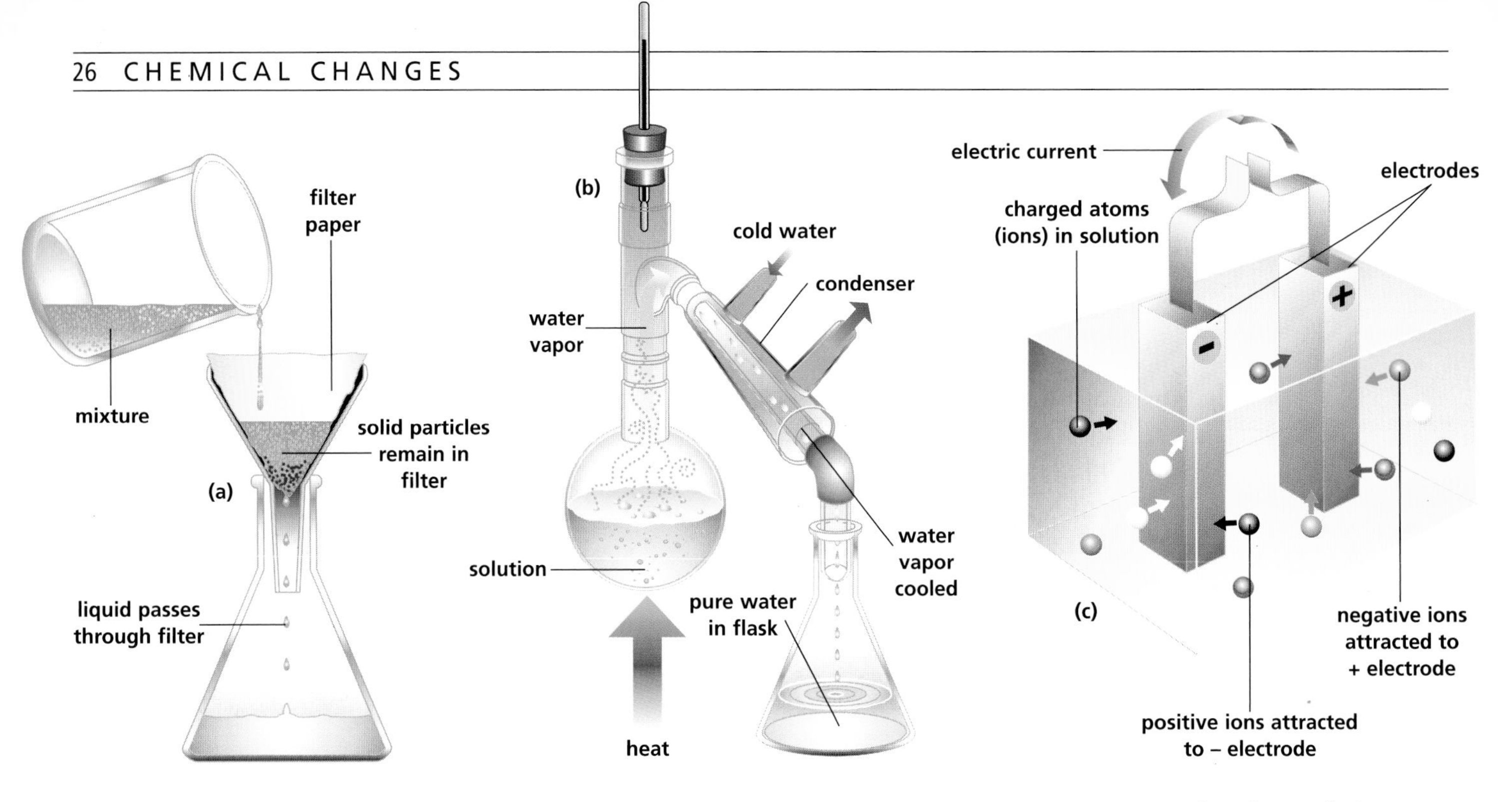

▲ Common processes used in chemical plants include mixing, dissolving, filtering (a), and distilling (b shows distilling water). Electrolysis (c) is a process in which electricity is used to separate chemicals.

Polymerization is the reaction by which plastics are produced. It involves the linking together of small molecules (monomers) to form large ones (polymers). Cracking is an operation in which the opposite happens—large molecules are broken down into smaller ones. It is an important reaction in refining petroleum.

▼ A chemical engineer inspects a small pilot plant to find out how a large chemical plant might work.

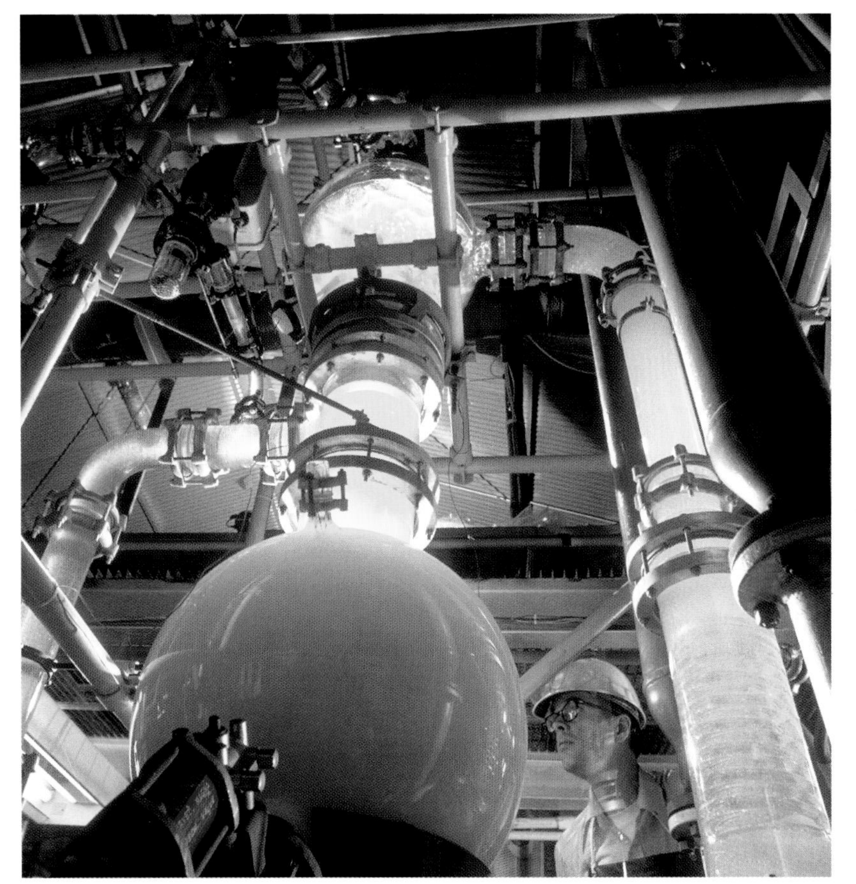

Many reactions have to take place at high temperatures or high pressures: the common plastic polyethylene is produced at pressures of up to 2,000 times normal atmospheric pressure. Other reactions will not take place properly unless a catalyst is present. A catalyst is a substance that can make a reaction go faster, but is not itself changed in the process.

Chemical engineering

The people who design the chemical plants (factories) in which the reactions are carried out are called chemical engineers. They have to take a chemical process that works in the laboratory and make it work on a large scale.

Chemical engineers have to design the vessels (containers) in which reactions take place. They must also design the equipment that ensures that the chemicals involved are in the right state and in the right place at the right time. This includes pumps, heaters, pipes, valves, and so on.

When designing a new chemical plant, the engineers first build a small-scale one, called a pilot plant. If this works, they can go ahead with the full-size plant.

DIRT-BUSTERS

At the bird sanctuary, a team of workers has brought in a flock of sea birds. They have been covered in oil leaking from a wrecked ship. The team sets to work cleaning the birds with detergent. This breaks up the oil into droplets, which are then washed away with water. The birds look bedraggled, but they will survive.

Detergents are powerful cleaning agents that will get rid of oil on sea birds, as well as grease and dirt on dishes, clothes, and so on. Detergents are synthetic products made from petroleum chemicals. Before people had detergents, the main cleaning agent was soap. Today we use soap mainly for cleaning ourselves.

key words

- detergent
- molecule
- soap

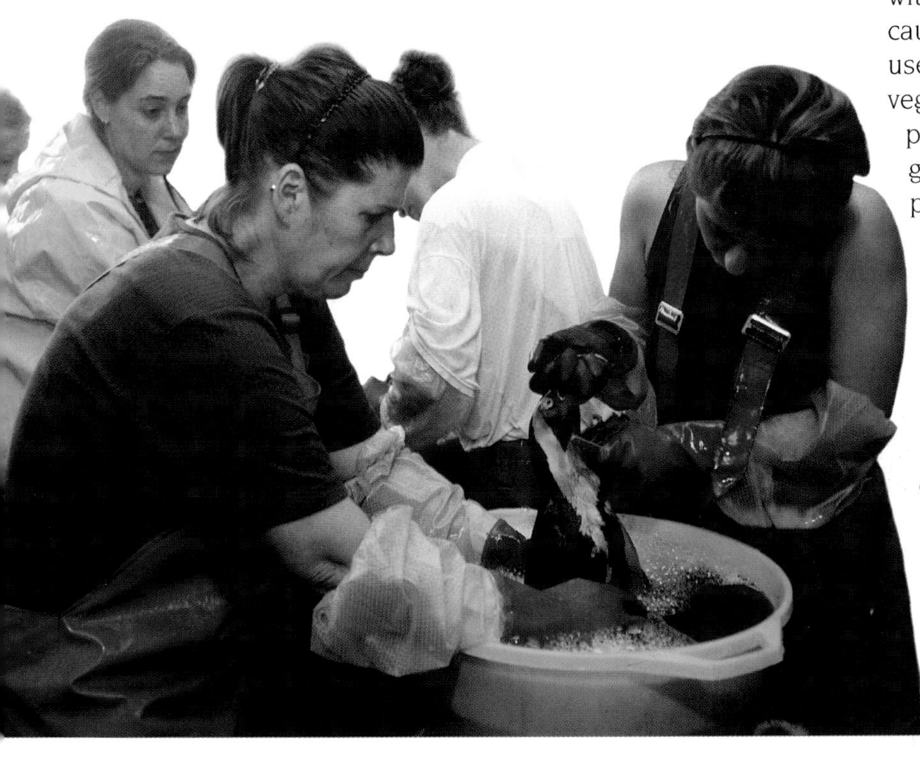

▼ Following an oil spill off the South African coast, teams of workers clean penguins by scrubbing them with detergent.

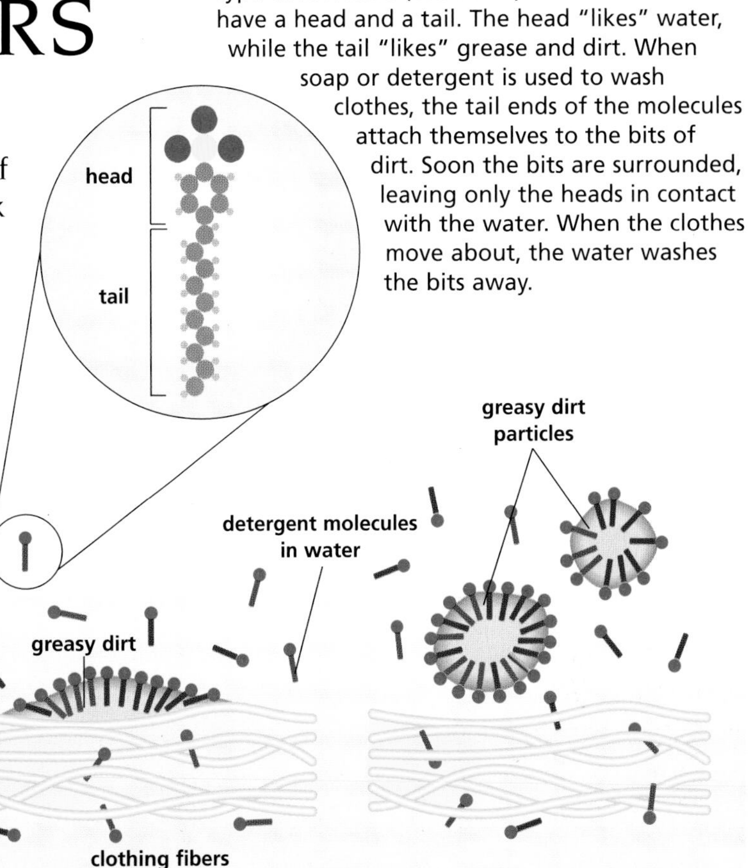

▼ Soaps and detergents are made up of a similar type of molecule (basic unit). These molecules have a head and a tail. The head "likes" water, while the tail "likes" grease and dirt. When soap or detergent is used to wash clothes, the tail ends of the molecules attach themselves to the bits of dirt. Soon the bits are surrounded, leaving only the heads in contact with the water. When the clothes move about, the water washes the bits away.

Soap has been made for thousands of years. It is produced by heating fat or oil with a substance that is an alkali, like caustic soda. In the past, animal fats were used, but now soaps are made using vegetable oils such as palm oil. Another product we get from making soap is glycerine, which can be used to make plastics and explosives.

Beating the scum

The problem with soap is that it sometimes forms a messy scum. Detergents do not. They also have a more efficient cleaning action than soap. Different detergents are made for different purposes. Dishwashing liquid, laundry detergents for washing clothes, and shampoos all contain different ingredients. Laundry detergents, for example, can contain brighteners to make fabrics look extra bright, and enzymes, which can dissolve stains such as sweat and blood.

Refining the Crude

The bar of the pole vault is set at 20 feet (6 m). The athlete sprints towards it, long pole in hand. He digs the pole into the ground and launches himself into the air. The pole bends right over, looking as though it will break, but then straightens. It helps throw the vaulter over the bar, with only an inch to spare.

No natural material, like wood or metal, could be used for such a vaulting pole. It would not be light enough, flexible enough, or strong enough. The pole is made of fiberglass, a plastic material that is reinforced (strengthened) by having long glass fibers running through it. Such a material is called a composite. Another type of composite uses carbon fibers to strengthen the plastic. It is used, for example, to make tennis rackets.

Composites and plastics are examples of synthetic materials. These are manufactured from chemicals made in factories, not from natural materials. Thousands of different products are synthesized from chemicals these days—not only plastics, but also dyes, drugs, detergents, explosives, fibers, pesticides, and so on.

Sources of synthetics

Most synthetic materials are manufactured from organic chemicals, which are carbon compounds. These compounds are called "organic" because it was once thought that they could only be produced by living things (we now know this is not true).

▲ Oil refineries are run mainly from the control room, where workers can monitor the various processes. Everything works automatically, under computer control.

◀ In an oil refinery, crude oil is processed into many different products, including gasoline, kerosene, and diesel oil.

key words

- cracking
- distillation
- fraction
- hydrocarbon
- molecule
- plastics
- refining

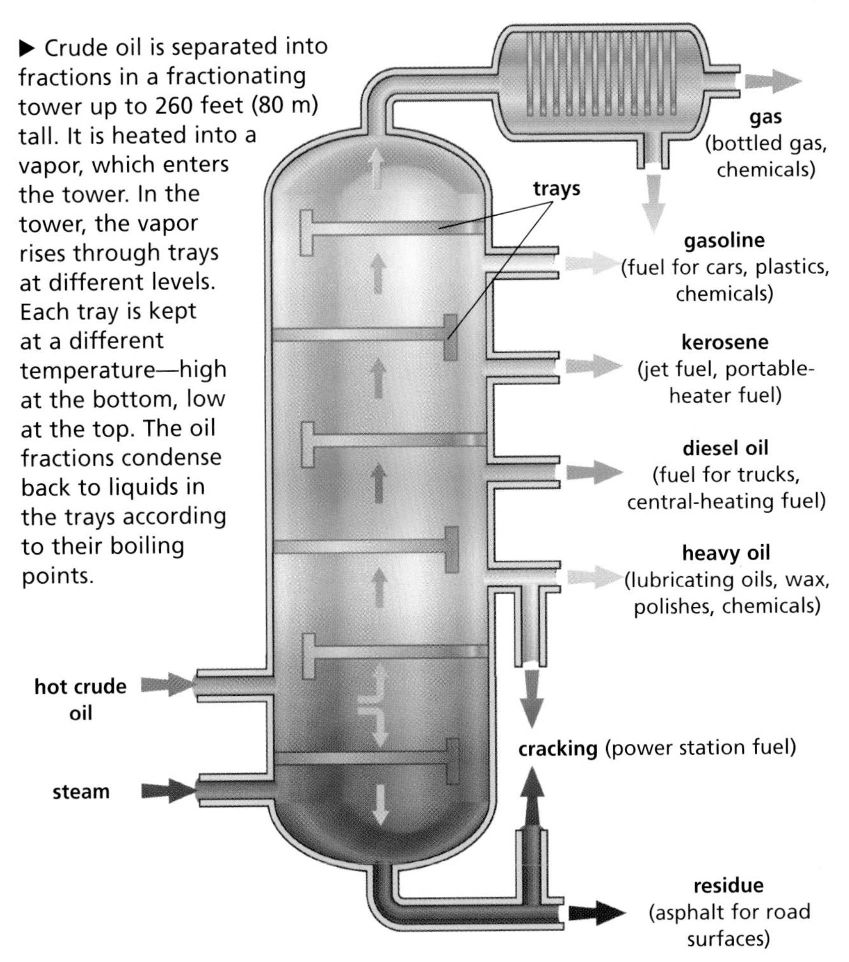

▶ Crude oil is separated into fractions in a fractionating tower up to 260 feet (80 m) tall. It is heated into a vapor, which enters the tower. In the tower, the vapor rises through trays at different levels. Each tray is kept at a different temperature—high at the bottom, low at the top. The oil fractions condense back to liquids in the trays according to their boiling points.

Living things contain thousands of different carbon compounds, and it is from the remains of living things that we get our organic chemicals. We get them mostly from crude oil (petroleum), which scientists believe is the remains of tiny creatures that died and decayed millions of years ago.

Crude oil is a mixture of hundreds of different carbon compounds. They are called hydrocarbons because they are made up of carbon and hydrogen. Their molecules (basic units) have a backbone of carbon atoms joined together in chains or rings.

Sorting out

The hydrocarbon mixture that is crude oil is of little use as it is. The hydrocarbons have to be sorted out before they become useful. Fortunately, this is quite easy because they all have different boiling points. This means that they can be separated by distillation. In this process, a substance is heated so that it evaporates (turns to vapor). Then the vapor is allowed to cool and condense (turn back into liquid).

Distillation is just the first stage in processing, or refining, the oil. It separates the oil into various parts (fractions). Some of these fractions can be used directly as fuels. Other fractions and gases provide the starting point for making a wide range of chemicals, often called petrochemicals.

Get cracking

Heavy oil fractions can be made into more useful products by cracking. This is a process by which the large molecules in the heavy oil are broken down into smaller ones. This produces lighter oils, fuels, and gases. In turn, the gases can be converted into fuels and chemicals. This is done by polymerization, a process of building up larger molecules from smaller ones. Together, these and the many other processes that take place in an oil refinery, make it possible to use every part of the crude oil.

▼ Crude oil is such a valuable source of chemicals that it seems almost a waste to burn it as fuel. A tank of gasoline takes an average family car about 300 miles (500 km). But we could convert that fuel into chemicals, which could be used to make any of the products shown here.

10 polyester shirts

260 feet (80 m) of water pipes

5 plastic crates

15 rolls of nylon twine

3 plastic trash cans

2 car tires

PLASTICS GALORE

Chemists make plastics in much the same way as we make daisy-chains. They link together chemicals with short molecules (basic units) in chains to make long molecules. If a short molecule was the length of a daisy stalk, a plastics molecule would be up to 9,800 feet (3,000 m) long.

CELLULOID AND BAKELITE

A U.S. inventor named John Hyatt (1837–1920) made the first successful plastic in 1870, when he was trying to make an artificial ivory for billiard balls. His invention was made from the cellulose in wood and was called celluloid. The modern plastics industry was born in 1909, when the U.S. chemist Leo Baekeland (1863–1944) made Bakelite while trying to produce new varnishes. Bakelite was the world's first synthetic plastic.

John Hyatt

Leo Baekeland

All the materials we call plastics are made up of long molecules. It is these long molecules that make plastics so special. Most other materials have short molecules, made up of just a few atoms joined together. In a plastics molecule, tens of thousands of atoms may be joined together.

(a)

(b)

▲ Kevlar is a tough plastic that is closely related to nylon. It is used for making bullet-proof vests. Kevlar molecules are long polymer chains (a), made up of individual units (b) that contain two carbon rings.

◀ A technician in a plastics factory checks the production of plastic tubing. Wide tubing like this is used to make trash bags and plastic sheets.

key words

- molecule
- molding
- polymer
- polymerization
- synthetic
- thermoplastic
- thermoset

There are substances in nature that have long molecules, including rubber and wood. But we do not call them plastics. Plastics are synthetic substances, manufactured from chemicals. An important property of many plastics is that they are easy to shape by heating.

We find plastics everywhere. We drink from plastic cups, fry food in nonstick plastic-coated pans, wrap goods in plastic bags, and wear plastics in the form of synthetic fibers. Drainpipes, squeeze bottles, heatproof surfaces, floor coverings, tires, and superglues are just a few of the many other products made from plastics that we come across every day.

Many parts

Another general name for plastics is "polymers." The word means "many parts." This tells us that plastics are made by stringing many small parts (short molecules) together. The chemical process of making plastics is called polymerization.

◀ This tailplane for the Bombardier Global Express jet aircraft is made from a light, strong composite material. A composite has fibers of a very strong material such as carbon fiber or Kevlar embedded in plastic.

In almost all plastics, the long chains that form the molecules are linked together by carbon atoms. Carbon is the only chemical element that can link together in this way.

Thermoplastics and thermosets

All plastics are shaped by heating, but react differently to heat afterward. If you place a hot saucepan on the plastic countertop in the kitchen, nothing happens. But place it in an empty polyethylene dishwashing pan, and the plastic will melt. This happens because the two plastics are different types.

Polyethylene softens and melts when heated. It is a type of plastic called a thermoplastic. Many other common plastics are thermoplastics, including PVC (polyvinyl chloride), polystyrene, and nylon.

The countertop plastic does not soften or melt when heated, although it will burn eventually. It is a type of plastic called a thermoset. The countertop is made from a thermoset called melamine-formaldehyde, after the substances it is made from. The first synthetic plastic, Bakelite, was made from phenol and formaldehyde.

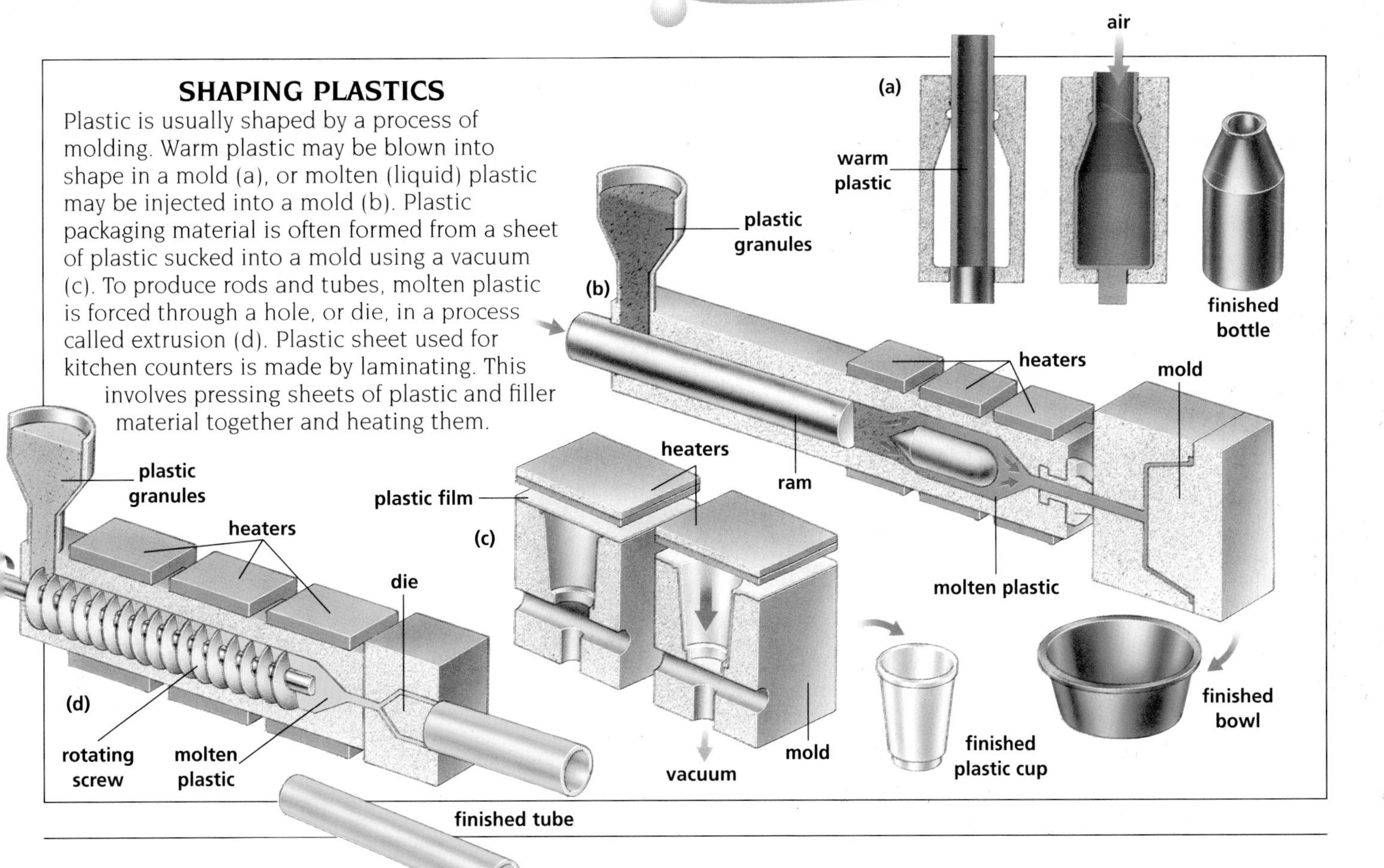

SHAPING PLASTICS

Plastic is usually shaped by a process of molding. Warm plastic may be blown into shape in a mold (a), or molten (liquid) plastic may be injected into a mold (b). Plastic packaging material is often formed from a sheet of plastic sucked into a mold using a vacuum (c). To produce rods and tubes, molten plastic is forced through a hole, or die, in a process called extrusion (d). Plastic sheet used for kitchen counters is made by laminating. This involves pressing sheets of plastic and filler material together and heating them.

CREATING COLORS

Buddhist priests the world over wear bright orange robes, which get their color from the dye saffron. This dye is obtained from the stigmas (pollen-collecting organs) of crocus flowers. The stigmas of over 150,000 blooms are needed to make 2.2 pounds (1 kg) of saffron dye.

▶ Newly dyed cloth hangs out to dry in the streets of Marrakesh, Morocco.

Saffron is one of several natural dyes that have been used to color fabrics since ancient times. Other plant dyes include madder (red) and indigo (blue), which come from the madder and indigo plants. Natural dyes can also come from animals. Cochineal red, for example, is a dye extracted from insects.

Most dyes used today, however, are synthetic. They are usually made from chemicals obtained from crude oil. Aniline, for instance, is a compound from oil that is the starting point for making many dyes.

The molecules that make up most dyes, whether natural or synthetic, contain rings of carbon atoms. These carbon rings are important in giving a dye its color.

key words

- dye
- mordant
- pigment
- ring compound
- synthetic

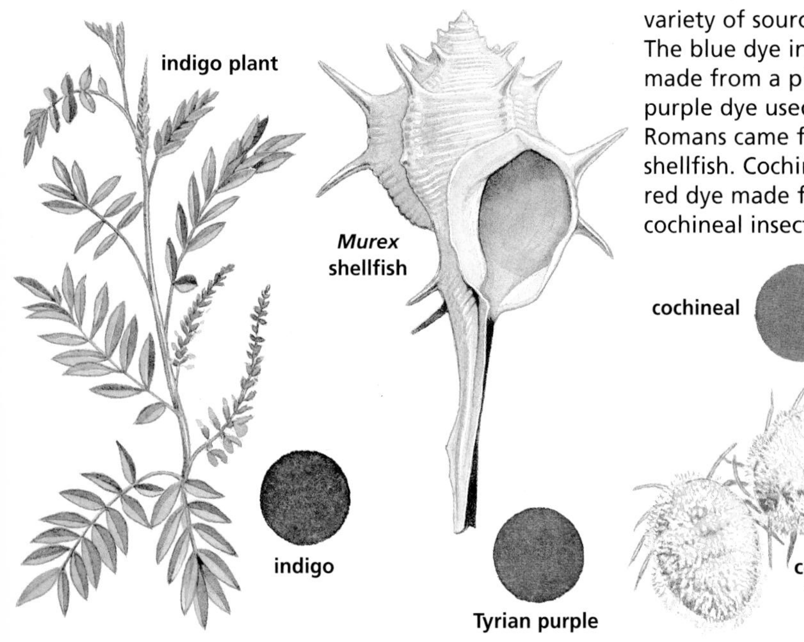

▼ Traditional dyes come from a wide variety of sources. The blue dye indigo is made from a plant; a purple dye used by the Romans came from a shellfish. Cochineal is a red dye made from cochineal insects.

Fast colors

In general, synthetic dyes have more brilliant colors than natural dyes. They bond well with fabric fibers so that they do not readily wash out. Also, they are colorfast, which means they do not fade easily.

Natural dyes do not bond well with natural fibers such as cotton. These fibers first have to be treated with a substance called a mordant. The mordant clings to the fibers and then bonds with the dye.

Pigments

Dyes are coloring substances that dissolve in water. Pigments are coloring substances that do not dissolve in water or other solvents (dissolving liquids). They are used to color materials such as inks, paints, and cosmetics. Traditional pigments include colored earths such as ocher, and metal compounds such as iron and titanium oxides. Most pigments are now synthetic.

WILLIAM PERKIN

The English chemist William Perkin (1838–1907) discovered the first synthetic dye, mauvein, in 1856, and set up a factory to produce it. He made it from aniline, a chemical now extracted from oil.

THIN PROTECTION

Left to themselves, wood rots and iron rusts when exposed to the weather. But give them a thin coat of paint, and they can last for centuries.

Paint is a liquid mixture containing coloring matter, which dries to form a thin, tough film. This film decorates the surface underneath and protects it from attack by water, air, and insects.

The substance that forms the film is called the vehicle, or binder. The coloring is provided by pigments. To make the mixture flow easily when applied, it is dissolved in a solvent, or thinner.

▶ Wearing a mask as protection against poisonous fumes, a worker spray-paints the side of a car.

▼ Many coatings are needed to build up the shiny paintwork on a car body. First, the body is dipped in phosphate solution to help it resist corrosion. Then it is given one or two coats of primer, followed by a base coat, an undercoat and one or more top coats. The final coats are baked, or stoved. (The colors shown are not the actual colors of different coatings.)

key words

- emulsion
- evaporation
- pigment
- solvent
- vehicle

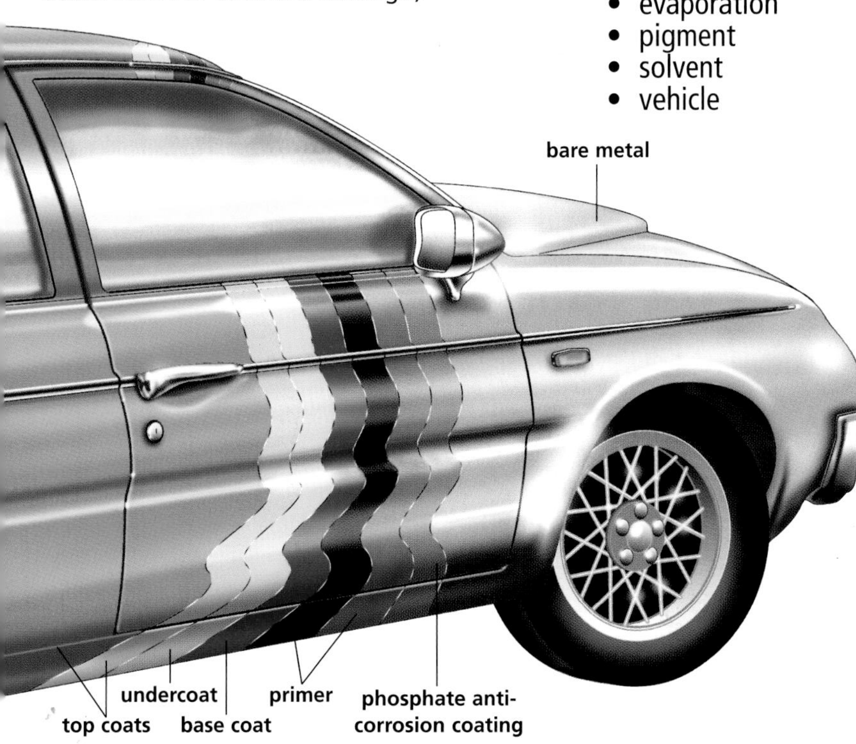

Early paints used natural oils like linseed oil as a vehicle. Modern paints use synthetic resins, which are kinds of plastics. In the oil paints generally used for painting outside, the usual solvent is petroleum spirits, a product of oil refining.

A strong bond

After the paint has been applied, two processes take place. The solvent evaporates, and the binder reacts with the oxygen in the air. As a result, the molecules of the binder join up to form a rigid polymer, or plastic material—the paint film.

Many paints these days use water as a solvent, and so are cleaner to use and cause no air pollution—unlike some oil-based paints. One kind of water-based paint is an emulsion. In an emulsion, the vehicle particles do not dissolve, but are spread out in the water as very fine droplets.

A clear, hard varnish called shellac is made from the secretions of the tiny red lac insect from Thailand. At certain times of year, lac insects swarm in huge numbers on trees, literally feeding themselves to death. But before she dies, each female lays her eggs on the tree, protecting them with a clear material called lac. This is later harvested and made into shellac.

DEALING WITH DISEASE

When the native peoples who live in the Amazon jungle get a fever, they chew the bark of the cinchona tree. The fever is brought on by malaria. Cinchona bark contains a drug, quinine, that fights the disease.

▲ This enlarged picture shows the effect of an antibiotic on bacteria. The bacterium on the right has not yet been damaged, but that on the left has been attacked and destroyed.

Malaria is one of the most widespread diseases in the world, affecting as many as 500 million people. Quinine is still one of the main drugs used to treat it. But nowadays quinine is a synthetic product, manufactured from chemicals.

Most drugs, also called pharmaceuticals, are now synthetic. But quite a few natural drugs are still used. One long-used plant drug, digitalis, is extracted from foxglove. It is used to treat heart conditions.

Another natural drug is morphine, extracted from the seeds of the opium poppy. It is a powerful painkiller. Heroin comes from the same plant. Both morphine and heroin are highly addictive, or habit-forming. The use of heroin is one of the major causes of drug addiction.

PENICILLIN PIONEERS

Alexander Fleming (1881–1955) discovered penicillin in 1928 while working at St. Mary's Hospital, London. But it was not produced in a pure form until 1940, when it was shown to have amazing antibiotic properties. This latter work was carried out at Oxford by a team led by Ernst Chain (1906–1979) and Howard Florey (1898–1968). The three pioneering scientists shared the 1945 Nobel prize for medicine.

Ernst Chain

Howard Florey

◀ Certain drugs are illegal and people try to smuggle them from country to country in their luggage. Police use specially trained dogs to sniff out these hidden drugs.

key words

- antibiotic
- bacteria
- drug addiction
- vaccination
- virus

Beating bacteria

Tiny, microscopic organisms called bacteria are the cause of most dangerous diseases, including blood poisoning, cholera, typhoid, and tuberculosis. Bacterial infection can be treated by synthetic drugs such as the sulfonamides, but these days it is usually treated by antibiotics.

Antibiotics are substances produced naturally by certain molds and bacteria. Penicillin was the original antibiotic and is still the most widely used. Others include streptomycin, Terramycin, and tetracycline. Each of these antibiotics is suited to treating certain diseases.

The use of antibiotics over the years has dramatically reduced the death rate from disease. But antibiotics have been used so much that some bacteria are becoming resistant to them—which means that the antibiotic cannot destroy the bacteria. So scientists are always looking for new antibiotics.

Viruses and vaccines

Antibiotics cannot cure all diseases. In particular, they cannot treat diseases caused by viruses, such as influenza, measles, mumps, hepatitis, and polio.

But doctors can prevent some virus diseases by vaccination. This involves injecting a vaccine into the patient's body. A vaccine is made up of a dead or weakened form of the same virus. The body produces antibodies to fight the invading virus. Later, if the body is exposed to the real virus, the antibodies are already there to attack it before it multiplies.

The English doctor Edward Jenner pioneered vaccination to treat smallpox in 1796. A worldwide mass vaccination program 200 years later wiped the disease from the face of the Earth.

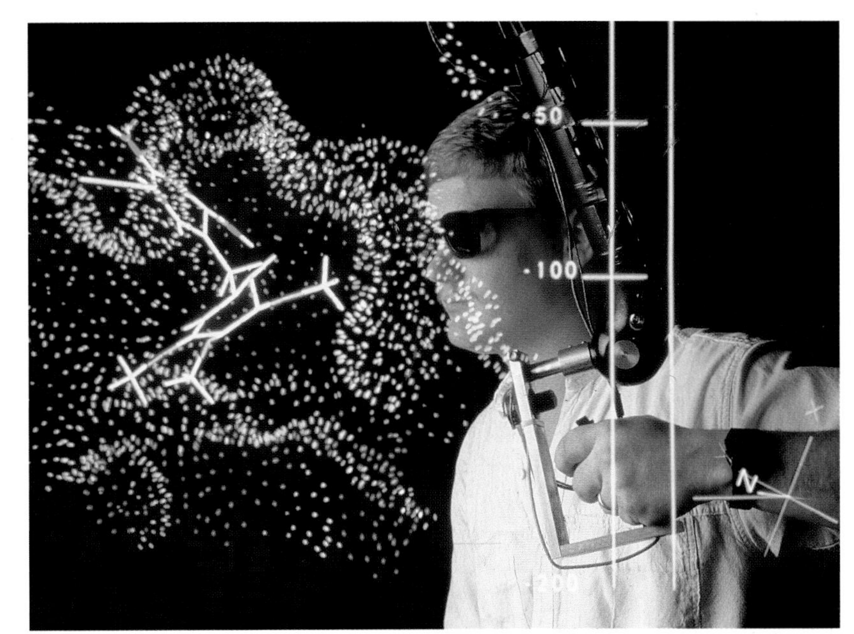

▲ This chemist is using a virtual reality program to look at the action of a drug molecule.

Designer drugs

Some synthetic drugs, such as quinine, are exact copies of natural substances.

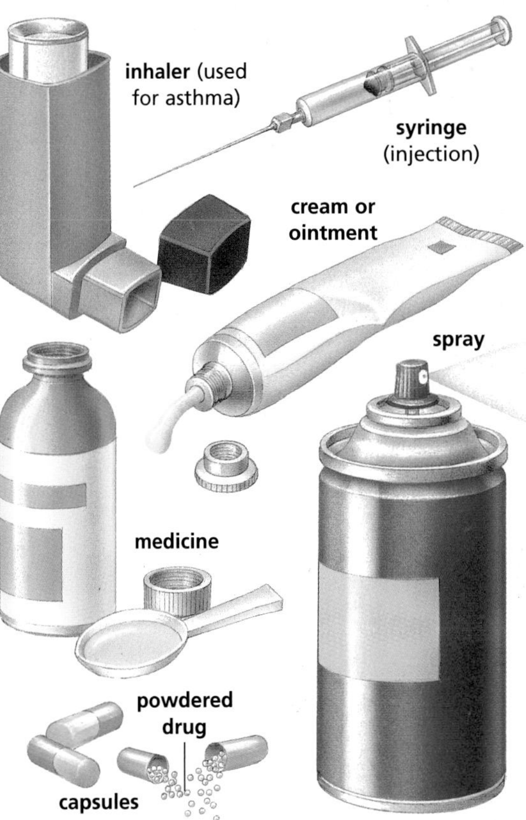

▶ Different ways of administering (giving) drugs.

COMMON TYPES OF DRUG	
Type	**Effects**
Anesthetic	Prevents patients feeling pain; local anesthetic acts locally; general anesthetic creates unconsciousness
Analgesic	Prevents or reduces pain
Antibiotic	Kills the bacteria that cause disease
Antihistamine	Relieves symptoms of asthma, hay fever, and other allergies
Hormone	Used to overcome a hormone deficiency in the body
Narcotic	Helps prevent pain by deadening the whole nervous system
Sedative	Helps induce sleep
Tranquillizer	Helps calm a person
Vaccine	Helps the body fight a virus disease by triggering its natural defenses in advance

Increasingly, though, new drugs are being created to target diseases. Biochemists, who design drugs, can test a drug using a process called molecular modeling. This involves drawing models of the molecules (basic units) that make up the drug and the disease, on a very powerful computer. They then see how the different molecules behave together. This tells them whether or not the drug will be able to fight the disease.

FERTILE GROUND

In some countries, autumn can be a smelly time of year because many farmers pump sewage sludge onto their land. This sludge contains nitrogen compounds, which get washed into the soil and will help the next crop of plants to grow well.

▶ Manure is scattered over a field by a mechanical spreader.

Sewage sludge is one of several materials farmers spread on their land to help make the soil more productive, or more fertile. It is a fertilizer.

As plants grow, they take in certain essential elements from the soil. Fertilizers are designed to put back these elements so that future crops can also grow well. The most important elements are nitrogen, phosphorus, and potassium.

key words

- calcium
- manure
- nitrogen
- phosphorus
- potassium

Nitrogen fertilizers

Sewage sludge is just one way of replacing nitrogen. Farmyard manure is another. But most fertilizers are synthetic, produced from chemicals. These fertilizers include ammonia, and two chemicals made from ammonia—ammonium nitrate and urea. Ammonia production is a big part of the chemical industry.

Some plants make their own fertilizer. They are the legumes, and include crops such as beans and clover. Sometimes farmers grow these crops and then plow them into the soil as fertilizer. This process is called green manuring.

From rocks and bones

The chemical industry makes huge quantities of fertilizer that contains phosphorus. Called superphosphate, it is manufactured by treating phosphate rock with sulfuric acid.

Bonemeal is a phosphorus fertilizer used widely by gardeners. It is made from animal bones, which consist mainly of calcium phosphate. The phosphorus is released from the bonemeal slowly.

Potassium fertilizers come mainly from mineral deposits. Often they are mixed with nitrogen and phosphorus compounds to form a compound fertilizer.

▼ The effects of growing plants with different amounts of nitrogen fertilizer ("N" on the labels). The plant on the left has added sulfur (S) but no fertilizer.

STICKING TOGETHER

When you lick a stamp, join the parts of a model together, or stick a Post-it® note in a book, you are using adhesives. Adhesives are sticky substances that bond surfaces together.

Glues and gums are adhesives made from natural materials. Glues come from animals. They are made by boiling up such things as the bones and skin of cattle, and fish bones. Gums come from the sticky resins made by certain plants.

However, most adhesives used these days are synthetic, and are usually made from petroleum chemicals. These adhesives are plastic materials, or polymers, which set after they have been applied.

Making contact

There are hundreds of different kinds of synthetic adhesives. For example, contact adhesives are made from synthetic rubber dissolved in a solvent. You coat each of the surfaces to be joined with adhesive, let them dry for a while, and then press the surfaces together.

Model-makers use an adhesive containing the plastic polystyrene in a solvent. After it has set, it can be softened by warming, which makes it easy to reshape joints. It is called a thermoplastic adhesive, because it softens when heated.

key words

- adhesive
- glue
- resin
- synthetic
- superglue
- thermoplastic

(a) In the tube, a substance in the superglue called a stabilizer stops the molecules from linking together.

(b) When the glue is spread on a surface, minute traces of water stop the stabilizer from working.

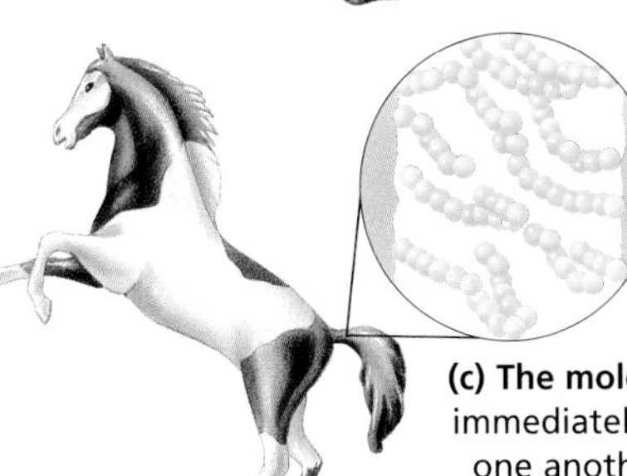

(c) The molecules of resin immediately link up with one another to form a solid adhesive.

▶ How superglue works

Big sticks

Among the strongest adhesives are the epoxy resins. They have even been used to stick cars to advertising billboards. Epoxy resins often come in two parts that have to be mixed together. One part is the resin, the other is a hardener. Adding the hardener makes the resin set into a rigid plastic material, or thermoset polymer. This process takes about half an hour. The adhesives called superglues, however, set in seconds. They are made from acrylic resins.

▼ High-tech aircraft like this B-2 Spirit stealth bomber are made from a number of different extralight materials. Superstrong adhesives are often used to join these materials together.

BIG BANGS

ALFRED NOBEL
The Swedish chemist Alfred Nobel (1833–1896) invented dynamite in 1867. Feeling guilty that his invention caused so much death and destruction, he set up a fund to award the annual Nobel prizes, one of which was to be a peace prize.

"Whoosh!" goes the firework rocket as it shoots high into the sky. The colored stars it releases explode with bangs like pistol shots. The substance that propels the rockets and makes the bangs is gunpowder, an explosive the Chinese first made over 1,000 years ago.

Gunpowder got its name because it was once used to fire bullets from guns. It is a mixture of chemicals—carbon (in the form of charcoal), sulfur, and potassium nitrate. When ignited (lighted), these substances burn rapidly and produce large amounts of gases. These gases expand suddenly, in a violent explosion. The explosion creates shock waves, which we hear as a bang.

▲ High explosives blast open a hillside in Arizona to make way for copper miners.

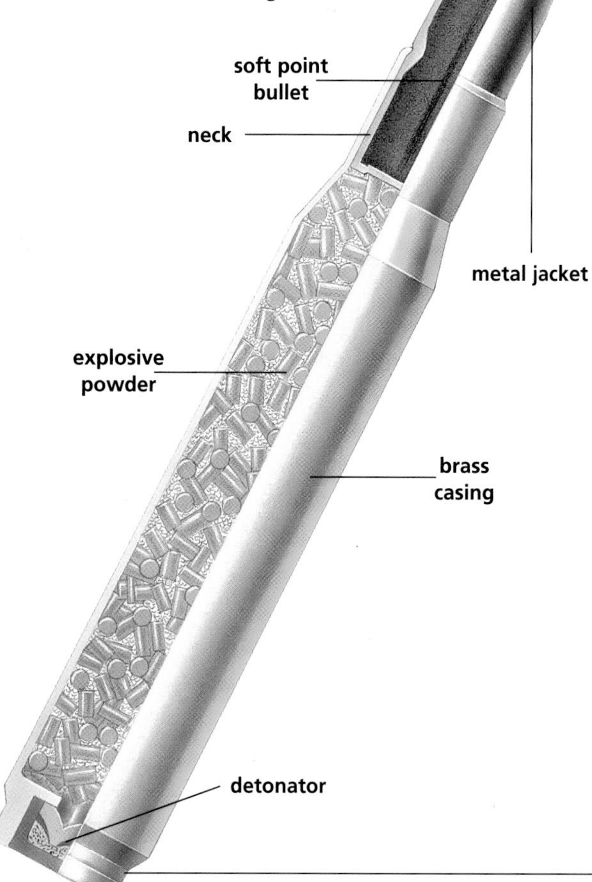

◀ The ammunition used in rifles is called a cartridge. When the rifle's firing pin strikes a detonator in the base of the cartridge, it ignites a low-explosive powder. This fires a metal bullet from the gun.

key words
- detonator
- high explosive
- low explosive
- nitrogen

All explosives work in this way. Some, called high explosives, burn many thousands of times faster than gunpowder. They are used in mining and tunneling to blast rocks apart, and in weapons such as shells and bombs.

The nitrogen connection

Almost all explosives contain nitrogen. The nitrogen compound ammonia is the starting point for several explosives.

Two of the most powerful high explosives are nitroglycerine and TNT (trinitrotoluene). Nitroglycerine is an oily liquid that is very dangerous to handle, because it explodes easily. In dynamite, nitroglycerine is mixed with an earthy material to make it safer to use.

All explosives need something to set them off. Low explosives can be set off by a burning fuse or by a sharp blow. High explosives have to be set off by a detonator.

CONTROLLING PESTS

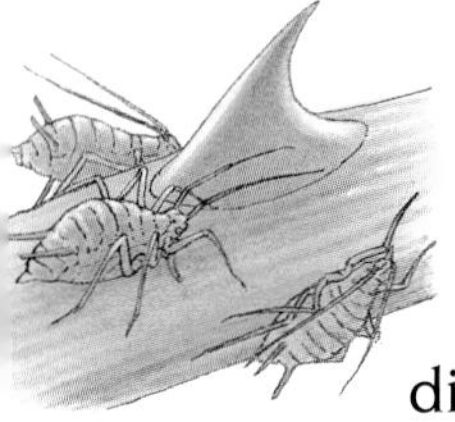

The stem of the beautiful rose is covered with aphids feeding on the sap, while ladybirds and lacewings feed on them. The gardener sprays the rose with insecticide and soon the aphids die. But the ladybirds and lacewings fly off—they are not affected.

Insecticides are one kind of pesticide, a chemical that kills harmful organisms on farms and in homes and gardens. The other main types of pesticide are herbicides, designed to kill weeds, and fungicides, which kill fungus diseases.

Some minerals and plant extracts are used as pesticides. Sulfur and copper sulfate, for example, are used to treat fungus diseases.

Most pesticides are synthetic, made mainly from petroleum chemicals. They are often compounds containing chlorine and phosphorus. The trouble is that they may kill not only harmful pests but also useful ones. They may also build up in the environment. This happened widely in the 1950s and 1960s, when a pesticide called DDT was used on a large scale. It was later found to be poisonous to birds and other animals.

▲ A crop-sprayer flies low to spray insecticide over a field.

key words

- environment
- fungicide
- herbicide
- pesticide

Being selective

DDT and similar substances have now been banned in many countries to protect the environment and wildlife. New pesticides have been developed that break down rapidly after they have been used and so are not a long-term danger.

Pesticides have also been developed that are selective in their action, like the insecticide that kills aphids but not ladybirds. Selective lawn herbicides kill broadleaf weeds but leave the thin blades of grass untouched.

Sometimes pests can be controlled without using chemicals. Parasites or other organisms that naturally prey on the pests are used. This method is known as biological control.

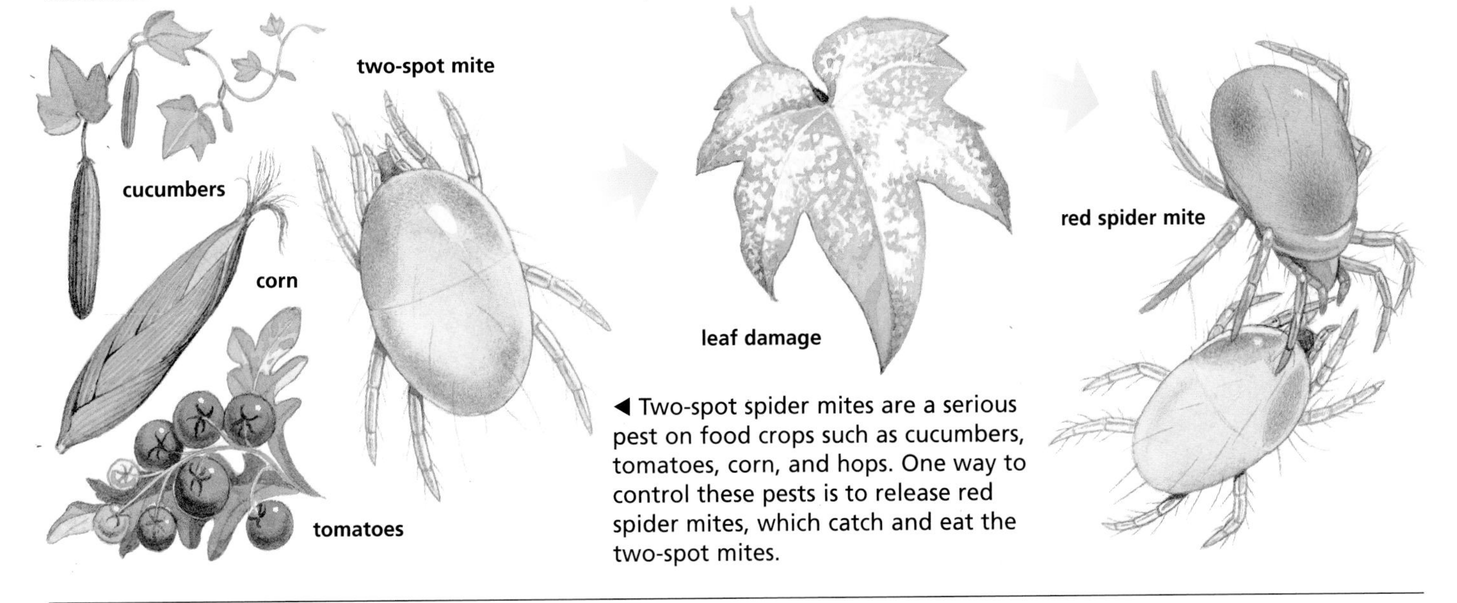

◀ Two-spot spider mites are a serious pest on food crops such as cucumbers, tomatoes, corn, and hops. One way to control these pests is to release red spider mites, which catch and eat the two-spot mites.

POISONING OUR WORLD

The year 2000 began badly for the environment. In January, oil smothered and killed over 150,000 sea birds on the northwest coast of France. In February, cyanide poisoned hundreds of tons of freshwater fish in Hungary's Tisa River. In March, raw sewage killed thousands of sea fish in a lagoon near Rio de Janeiro, Brazil.

Oil, deadly chemicals, and sewage are three of the things that can pollute or poison our environment and our water.

The oil that smothered the birds in France came from a wrecked oil tanker. At any time, hundreds of oil tankers are sailing the world's seas, together carrying millions of tons of oil. When tankers get holed in collisions, or run aground on rocks, vast amounts of oil can pour into the sea and drift ashore. All kinds of sea- and shore-life are affected, from shellfish to seals. Some animals may be saved by washing them with detergent, but most perish.

▶ This cormorant was just one of the many victims of a huge oil spill that took place off the Shetland Islands in 1993.

▼ The environment is under attack from all directions.

DISASTER AT BHOPAL

The world's worst chemical disaster happened in the Indian city of Bhopal in 1984. An explosion at a chemical plant making insecticides (below) released a cloud of deadly gas into the air. Because the gas was twice as heavy as air, it did not drift away, but formed a "blanket" over the surrounding area. It attacked people's lungs and affected their breathing. Eventually, as many as 3,000 people died and many thousands more had their health ruined.

Chemical attack

The cyanide that slaughtered the fish in Hungary came from a gold mine, where it was used to extract gold. Many industries produce poisonous waste products. Usually they are treated to make them harmless before they are released back into the environment.

Chemicals from farming also affect the environment. Farmers apply chemical fertilizers such as nitrates to the land. Sometimes they apply too much, and the surplus chemicals get washed by rain into rivers. Eventually, they can get into our drinking water.

More deadly than nitrates are the pesticides farmers use to protect their crops from insect and weeds. Many pesticides are persistent, which means that they remain active for a long time, and can be poisonous to other animals.

Polluting the air

The air can become polluted too. One of the main causes of air pollution is the car. When car engines burn their fuel, they give off fumes, such as nitrogen oxides and carbon dioxide, and soot. These fumes may cause breathing problems.

▲ The sun sets behind a blanket of smog hanging over Mexico City. Smog, a mixture of smoke and fog or chemical fumes, pollutes many of the world's big cities.

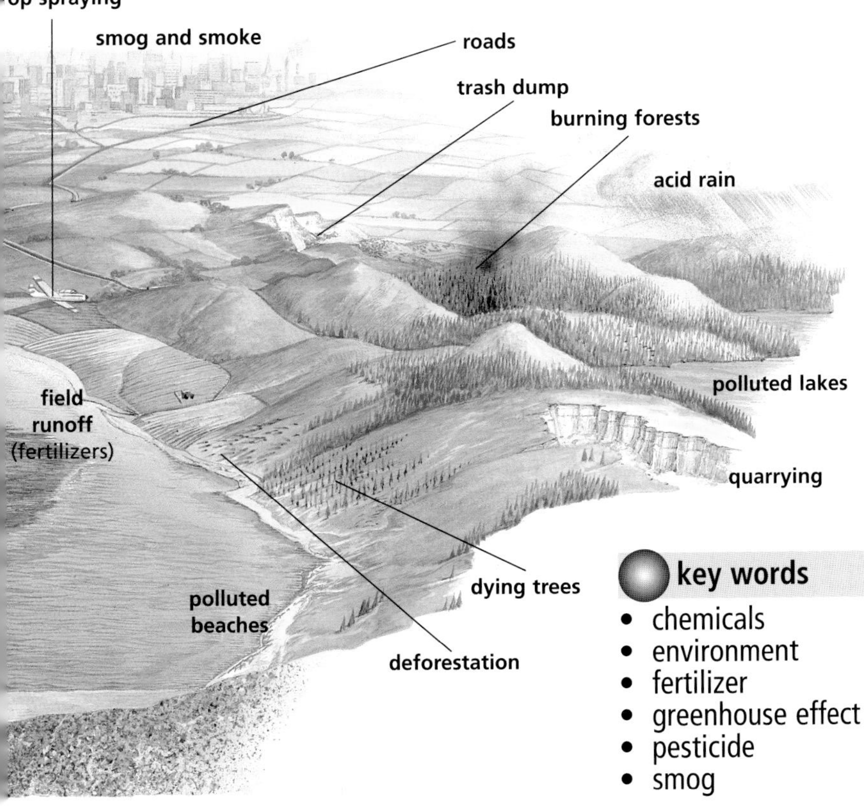

Raining acid

Power stations and factories burn fuels such as coal and oil to produce energy. They too give off fumes that pollute the air, including sulfur and nitrogen oxides. These oxides combine with oxygen and moisture in the air to form sulfuric and nitric acids. When it rains, the rain is acidic.

Acid rain falls into lakes and rivers, and can make them too acid to support plant and animal life. Acid rain also kills trees. Lifeless lakes and dying forests are already found in parts of northern North America and northern Europe.

In a greenhouse

The carbon dioxide that factories and cars produce when they burn their fuels builds up in the Earth's atmosphere. It makes the atmosphere act rather like a greenhouse and trap more of the Sun's heat. This is causing world temperatures to rise, or global warming.

Chemicals called chlorofluorocarbons (CFCs), found in sprays and refrigerators, increase the greenhouse effect. They also attack the layer of ozone in the Earth's upper atmosphere, which protects us from much harmful radiation from the Sun. If the ozone layer thins too much, it will let through more of this dangerous radiation.

key words

- chemicals
- environment
- fertilizer
- greenhouse effect
- pesticide
- smog

GETTING RID OF RUBBISH

Every day the people of New York City throw away up to 27,000 tons of waste—newspapers, plastic bags, food scraps, cans, old clothes, bottles, and so on. Until recently, twice a week this rubbish was collected and dumped on a site on Staten Island. It was the biggest garbage dump in the world.

▶ About 90 percent of the world's domestic waste goes to landfill sites.

Most household rubbish is disposed of at waste dumps like this. They are called landfill sites. Usually, a great pit is dug in the ground and filled with rubbish. The rubbish is then covered over and replanted with plants and grass.

The problem is that we are producing so much rubbish that we are running out of suitable sites. Also, harmful substances in the buried rubbish can find their way into our water supplies. And the rotting matter in the waste produces the gas methane, which can cause explosions. On many old landfill sites, this gas is piped away and used as fuel—it is the same gas that we use for cooking in our homes.

▶ At a sewage plant, screens and settling tanks first remove the larger objects, grit, and sludge. The sludge is further processed to make fuel gas and fertilizer. The water remaining is sprinkled over filter beds containing microbes that feed on any remaining waste. The water is then clean enough to go back into a river.

water returned to river

Up in smoke

Another method of dealing with rubbish is by burning, or incineration. Modern

◀ An engraving of the Paris sewers from the 1850s. Paris's sewers were neglected and little known until the early 1800s, when the city's works inspector, Pierre Bruneseau, mapped the 1,250-mile (2,100-km) network.

incinerators not only burn waste, but often use the heat this produces for heating or generating electricity. Before waste can be used as fuel, materials that will not burn, such as metal and glass, must be separated from those that will burn, like paper.

One disadvantage of burning wastes is that it can produce toxic (dangerous, poisonous) gases. Incinerators must therefore be fitted with efficient gas cleaners to remove these fumes.

Increasingly, local authorities are encouraging households to take their waste paper, metals, glass, and plastic to recycling centers. The different materials are then processed before being used again.

In the sewers

In additon to all the solid rubbish we throw away, we produce a lot of liquid waste. This includes body wastes from toilets, and water from sinks, washing machines, and baths. This liquid waste is called sewage.

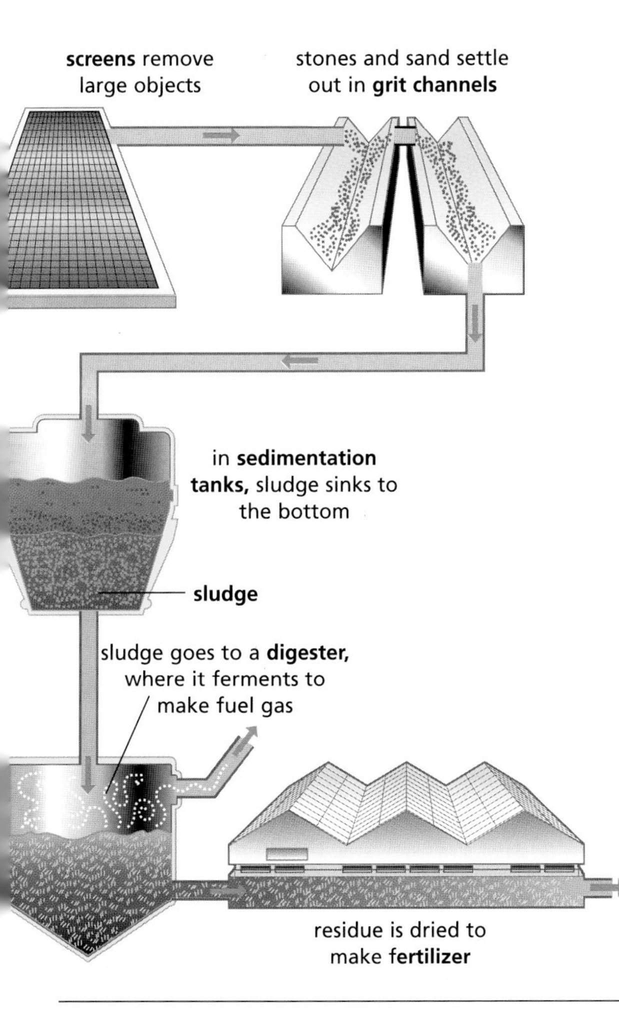

In some parts of the world, raw (untreated) sewage is pumped directly into rivers or the sea, polluting the water. But in most communities, a system of pipelines called the sewers carries the sewage to treatment plants. Here, solids and harmful substances are removed so that the water can be returned to the environment. The sludge left over can be treated to make fuel gas and fertilizer.

Wastes at work

Many industries, particularly the chemical industry, use dangerous substances. This produces wastes that can be harmful to our health and the environment. Landfill is often used for the safe disposal of solid wastes. Liquid wastes are treated to make them harmless before they are released. Unfortunately, this does not always happen, and pollution can result.

Some of the deadliest waste is produced by the nuclear industry. The waste is radioactive, which means that it gives off harmful radiation. Some waste remains active and dangerous for thousands of years.

For safety, most nuclear waste is stored underground. The surrounding soil and rock stop the radiation from reaching the surface. Liquid wastes are often kept in tanks surrounded by thick concrete. Some waste is made into a kind of glass and stored in sealed containers in deep mines.

key words

- incineration
- landfill
- pollution
- radioactive
- recycling
- sewage

▼ Liquid nuclear waste glass is poured into a steel mold. In glass form, nuclear waste can be disposed of safely.

REUSING RESOURCES

Next time you drink a can of cola or fizzy lemonade, spare a thought for the can. It is made of aluminium, and the metal has probably been used before. It might once have been cooking foil, bottle tops, or even part of an aircraft wing.

Aluminum is one of the materials that are recycled, or used again, in many countries. Glass, paper, and plastics are also widely recycled. Many towns have recycling centers, where people can deposit these materials. Elsewhere, there are scrapyards, which collect metal scrap, from old lead pipes to crashed cars.

There are three main reasons why recycling is a good idea. One, it reduces the amount of waste we produce. Getting rid of waste, from homes and industry, has become one of the world's biggest environmental problems. The American people alone throw away 220 million tons of rubbish every year. Most of this could be recycled.

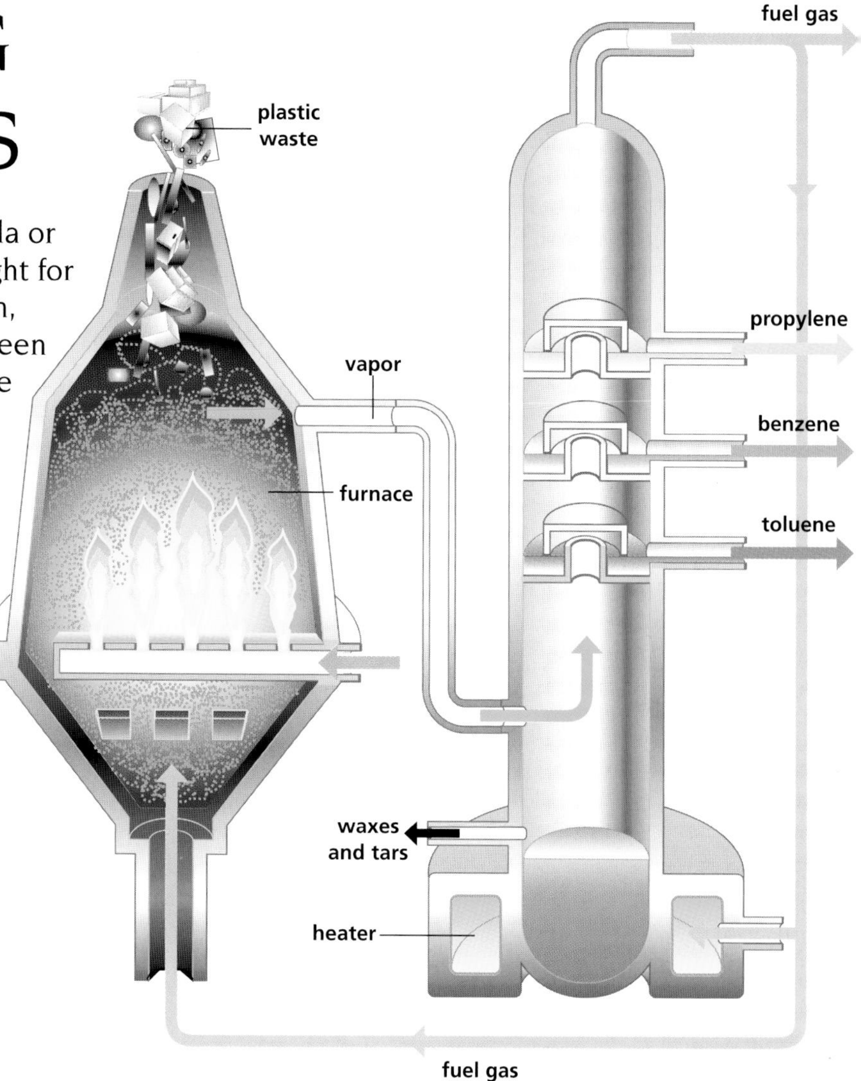

▲ In a recycling plant, plastics waste is heated at temperatures up to 1,500°F, in the absence of air. The materials turn to vapor, which then goes to a distillation tower. In the tower, the various substances in the vapor separate out at different levels. Some are used as fuels; others are valuable raw materials for the chemical and plastics industries.

Saving energy

Another reason for recycling is to save energy. This is why it pays to recycle aluminum. It takes large amounts of electricity to extract this metal from its ore, bauxite. Remelting aluminum waste takes much less energy and therefore saves money.

It also pays to recycle glass, for the same reason. Glass is made by melting together a mixture of sand and limestone. The mixture melts more easily when waste glass is added to it.

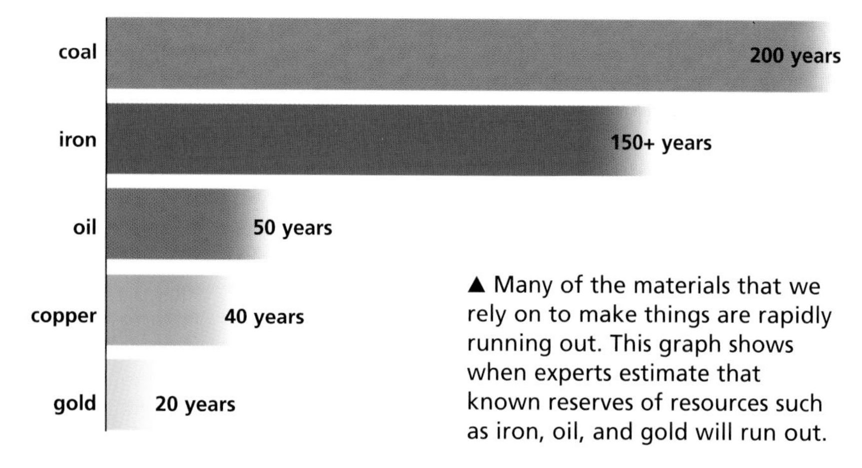

▲ Many of the materials that we rely on to make things are rapidly running out. This graph shows when experts estimate that known reserves of resources such as iron, oil, and gold will run out.

Thinking of the future

The third main reason for recycling is to save Earth's resources—the raw materials we need to make the goods we use. Paper, for example, is made from wood pulp, which comes from trees. The more paper we recycle, the fewer trees we have to cut down for wood pulp.

Trees are a renewable resource—we can keep growing them. Most of the materials we use, such as oil and minerals, are not. Once we have used them up, they will be gone for ever.

Oil, for example, will almost certainly run out later this century. This is why we need to recycle plastics, which are made from oil. Plastics can be broken down by heat into a range of chemicals, which can be used to make plastics once again. At present, however, recycling plastic is difficult and expensive.

▲ A digger lifts a pile of paper that has been shredded in preparation for recycling.

Precious metals

We get most of our metals by processing ores (minerals) that we take from the ground. The ores of some of our most useful metals, such as aluminum and iron, are found in huge quantities in the Earth's crust. They will last us for hundreds of years. But ores of other vital metals could run out later this century. These include lead, tin, zinc, copper, silver, and gold.

The loss of copper, silver, and gold would be particularly disastrous. Copper is vital to the electrical industry because it is one of the best conductors of electricity. Silver is essential to photography, as light-sensitive chemicals containing silver record the picture on film. Gold is used in jewelry, and increasingly in electronics.

key words

- biodegradable
- decomposer
- distillation
- oil
- ore
- recycling

The decomposers

Nature carries out some recycling for us. Organic waste, such as food scraps and vegetable matter, gets eaten by worms, insects, fungi, and bacteria. They convert it back into chemicals that enrich the ground and help new growth.

Substances that naturally rot away are biodegradable. Metals, glass, and most plastics cannot be broken down in this way. But biodegradable plastics are now being made. Some are designed to break down in sunlight. Others contain starch, which can be broken down by natural decomposers such as bacteria.

◀ Bales of crushed aluminum beverage cans, ready for recycling. Each bale contains more than 1 million cans.

GLOSSARY

This glossary gives simple explanations of difficult or specialized words that readers might be unfamiliar with. Words in *italics* have their own glossary entry.

alloy A mixture of *metals* or a metal and another substance, such as carbon.

antibiotic A drug, for example penicillin, that destroys bacteria or prevents them from growing.

atom The smallest particle of an *element*.

casting A method of shaping *metal*. Molten metal is poured into a mold and left to cool.

catalyst A chemical used to help bring about or speed up a chemical reaction, but which is not actually changed itself in the process.

ceramic A material made by baking earth substances such as clay at high temperatures.

composite A product made from *plastic* containing a strengthening mesh of fibers.

condensation The process by which a gas turns into a liquid.

conductor A substance, such as a *metal*, that allows an electric current to flow along it.

corrosion The gradual destruction of a *metal* by chemical action.

cracking A process carried out at an oil refinery in which large oil *molecules* are broken down into smaller ones.

electrolysis A method of extracting and purifying *metals* using electricity.

element A substance that is made of only one kind of *atom*. Elements cannot be broken down into other substances.

evaporation The process by which a liquid turns into a gas.

fertilizer A substance used to enrich the soil with the chemicals that plants need to grow well.

forging A method of shaping *metal* by hammering or squeezing.

fuel A material that is burned to produce energy. Coal, gasoline, heating oil, and natural gas are all fuels.

furnace A structure in which substances (such as *ores* and *metals*) are heated to very high temperatures.

insulator A substance, such as rubber, that blocks the flow of electric current.

metal An *element* that is usually shiny and solid at room temperature; metals are good *conductors* of heat and electricity. Metals that are found in the ground in metal form, for example copper, are called native metals.

mineral A natural solid material that has a specific chemical composition and a definite crystalline structure.

molecule A group of two or more *atoms* bonded to each other.

ore A *mineral* from which *metal* can be extracted.

pesticide A substance that is used, especially by farmers, to destroy insects or other animals that are harmful to plants or to animals.

petroleum Oil as it is extracted from the ground; also called crude oil.

plastic A *synthetic* substance with long *molecules* that is easy to shape; also called a *polymer*.

pollution The poisoning of the environment by such things as oil, chemicals, and car exhaust fumes.

polymerization A chemical process that makes large *molecules* out of small ones, producing polymers, or *plastics*.

recycling Using the same materials again, for either the same or a different purpose.

refining Purifying or separating materials (such as oil or *metals*) to produce more useful substances. See *cracking*.

refractory A material that withstands very high temperatures.

renewable resources Materials that, once used, can be replaced naturally, or by careful management. Wood is a resource that can be renewed by planting trees.

smelting The main method of extracting *metals* from their *ores*, by heating them in a *furnace*.

solvent A liquid that can dissolve another substance.

synthetic Made from chemicals.

vaccine A substance that is used to protect a person or animal from disease.

INDEX

Page numbers in **bold** mean that this is where you will find the most information on that subject. If both a heading and a page number are in bold, there is an article with that title. A page number in *italics* means that there is a picture of that subject. There may also be other information about the subject on the same page.

ACKNOWLEDGMENTS

Key
t = top; c = centre; b = bottom; r = right; l = left

Artwork
Baker, Julian: 18 b; 27 tr; 29 tl. **D'Achille, Gino:** 9 bl; 30 tr; 32 br; 34 cr; 38 tr. **Franklin, Mark:** 8 br; 10 b; 11 tr; 14 b; 31 b; 34 b. **Full Steam Ahead:** 25 tr; 44 bl. **Hinks, Gary:** 40–41 bc. **Howatson, Ian:** 10 tl. **Jakeway, Rob:** 26 t; 30–31 c; 42–43 bc; 44 tr. **Learoyd, Tracey:** 4 b; 19 cr. **Saunders, Michael:** 5 t; 13 tr; 16–17 tc; 21 bl; 22 b; 23 bl; 24 tr; 33 bl. **Smith, Guy :** 37 tr; 38 bl. **Sneddon, James:** 6 b. **Visscher, Peter:** 4 tl; 6 tl; 8 tl; 11 tl; 12 tl; 13 tl; 14 tl; 16 tl; 18 tl; 19 tl; 20 br; 21 tl; 22 tl; 23 tl; 25 tl; 27 tl; 28 tl; 29 br; 30 tl; 32 tl; 33 tl; 34 tl; 36 tl; 37 tl; 38 tl; 39 tl; 40 tl; 42 tl; 44 bl. **Woods, Michael:** 32 bl; 39 b.

Photos
The publishers would like to thank the following for permission to use their photographs.

Art Archive, The: 12 tr, bl (Archaeological Museum Zara/Dagli Orti); 13 bc (Araldo de Luca); 18 tr (Canterbury Cathedral/Dagli Orti); 42 bl (JFB).
Corbis: 11 bl (Charles E. Rotkin); 16 br (Araldo de Luca); 19 bl (Adam Woolfitt); 23 tr (James L. Amos); 26 bl (James L. Amos); 28 cl (Roger Wood); 32 tr (Robert Holmes); 34 bl (Galen Rowell); 36 tr (Gehl Company).
Digital Vision: 9 cr; 10 tr; 14 tr; 15 bl; 28 bl; 45 tr.
ISCOR Ltd.: 25 bl.
Levington Agriculture Ltd: 36 b.
Michelin: 22 tr.
Oxford Scientific Films: 7 bl (Peter Ryley); 20 t (Edward Parker); 38 cr (Marty Cordano); 39 cr (Tim Shepherd).
Panos Pictures: 7 cr (Chris Stowers); 40 bl (Roderick Johnson); 41 tr (Liba Taylor).
Photodisc: 37 b.
Science Photo Library: 4 tr (Bernhard Edmaier); 6–7 t (David Leah); 8–9 t (Rosenfeld Images Ltd.); 12 cr (Pascal Goetgheluck); 16 bl (John Howard); 17 cr (Lawrence Livermore National Laboratory); 21 tr (R. Maisonneuve); 23 br (Pascal Goetgheluck); 24 bl (Rosenfeld Images Ltd.); 30 bl (Pascal Goetgheluck); 33 tr (David Parker); 34 tr (CNRI); 35 tr (Peter Menzel); 40 tr (Simon Fraser); 42 tr (Simon Fraser/Northumbrian Environmental Management Ltd.); 43 br (US Department of Energy); 45 bl (Hank Morgan).
Scipix: 15 tr.
TRH Pictures: 31 tl (Short Brothers plc).
Woodfall Wild Images: 5 b (David Woodfall); 27 br (Heinrich Van de Berg).